Electrophilic halogenation

Reaction pathways involving attack by electrophilic halogens on unsaturated compounds

Cambridge Chemistry Texts

GENERAL EDITORS

D. T. Elmore
Professor of Biochemistry
The Queen's University of Belfast

A. Leadbetter
Professor of Physical Chemistry
University of Exeter

K. Schofield, D.Sc.
Professor of Organic Chemistry
University of Exeter

Electrophilic halogenation

Reaction pathways involving attack by
electrophilic halogens on
unsaturated compounds

Peter B. D. de la Mare

Professor of Chemistry
University of Auckland

Cambridge University Press

Cambridge

London · New York · Melbourne

CAMBRIDGE UNIVERSITY PRESS
Cambridge, New York, Melbourne, Madrid, Cape Town, Singapore,
São Paulo, Delhi, Dubai, Tokyo, Mexico City

Cambridge University Press
The Edinburgh Building, Cambridge CB2 8RU, UK

Published in the United States of America by Cambridge University Press, New York

www.cambridge.org
Information on this title: www.cambridge.org/9780521290142

First published 1976
Re-issued 2010

A catalogue record for this publication is available from the British Library

Library of Congress catalogue card number: 75-13451

ISBN 978-0-521-20968-7 Hardback
ISBN 978-0-521-29014-2 Paperback

Contents

v

Preface

The separation of the chemistry of aromatic systems from that of other unsaturated organic compounds is convenient for some purposes. It tends, however, to set up a number of artificial barriers; for example, between systems which 'should' substitute and those which 'should' add; and between reactions involving 'Wheland' intermediates and those involving carbocationic intermediates. The result too often constrains our attention into paths which become unsatisfactory, particularly when the reactions of halogens with unsaturated compounds are considered.

My interest in these processes was first stimulated by association with the late Professor P. W. Robertson, of Victoria University College in the University of New Zealand. This book, which is intended in part as a tribute to his inspiration, is an attempt to survey organic electrophilic halogenations and to illustrate the variety of ways in which carbocationic character can be developed in such reactions. General principles are emphasised; details are given only where necessary for illustration, and no attempt has been made to be exhaustive. Attention has been drawn wherever possible, directly or by inference, to the potential preparative significance of mechanistic findings. References are usually to recent articles, and are not intended to attribute priorities for ideas or for findings. Although many references are made to reviews, important recent work has necessitated rather extensive documentation in some areas.

I am immediately indebted to Professor R. C. Cambie and to Dr B. E. Swedlund for discussions of their recent work; and to Professor K. Schofield for his valued comments, criticisms and advice. This survey would have been much more imperfect without the help, encouragement and stimulation which I have had over the years from the many research workers and colleagues with whom I have been associated scientifically.

Mrs A. B. Bell typed the manuscript, and her help is much appreciated. I thank also the Syndics of the Cambridge University Press for the invitation to write this book, and the Council of the University of Auckland for refresher leave which made the undertaking possible.

P. B. D. DE LA MARE

Auckland, New Zealand, 1975

1 Terminology, definitions, methods of mechanistic study

'Every schoolboy knows...' (Macaulay)

1.1 Introduction

The halogens, of which fluorine, chlorine, bromine and iodine are the most familiar, form an important group of elements which exist not only in their well-known diatomic molecular forms (e.g. Cl_2), but also as atoms, as ions, and in covalent combination with many other elements. Their reactions with unsaturated organic compounds (that is, with compounds which possess a multiple bond to carbon) are encountered early in the study of organic chemistry; thus (1.1) represents the reaction of ethylene with chlorine, and (1.2) that of benzene with bromine.

$$CH_2:CH_2 + Cl_2 \longrightarrow Cl.CH_2.CH_2.Cl \qquad (1.1)$$

$$C_6H_6 + Br_2 \longrightarrow C_6H_5Br + HBr \qquad (1.2)$$

Historically, these reactions and their analogues have fascinated chemists for more than a century; they have been used to distinguish between olefinic unsaturation (as in ethylene) and aromatic unsaturation (as in benzene), and in this way and in many others have played a part in the development of some of the most important ideas of chemical theory. Currently, the mechanistic considerations which, as we shall see, enable the quite different reactions of (1.1) and (1.2) to be associated and interpreted within the framework of the chemistry of carbocationic intermediates put these reactions at the centre of development of organic chemistry. Practical considerations likewise require emphasis on the importance of an understanding of the courses taken in halogenations; the introduction of halogen at a particular point in an organic molecule can often be followed by its replacement by other functional groups or by other changes directed by the introduced halogen, so that the original point of entry provides a locus for subsequent synthetic processes.

In this book, we shall be concerned mainly with reactions of the

1

halogen molecules acting as *electrophiles*: that is, as seekers of electrons, or in other words as sources of supply of positive halogen for co-ordination with an electron-rich centre. Other related molecules which can behave similarly (e.g. chlorine acetate, $Cl.OAc$) will also be considered, as will the positively charged halogen ions.

The organic molecules to be discussed will be those which are formally unsaturated. Olefinic compounds, containing $C=C$ double bonds; acetylenic compounds, containing $C\equiv C$ triple bonds; aromatic compounds, containing cyclic conjugated systems of double bonds; systems involving double bonds between carbon and another element, including heterocyclic compounds; and compounds which are converted by reversible isomerism (*tautomerism*) into structures of one or other of the above kinds, will be among the important types to be considered.

The terminology to be adopted will essentially follow the conventions used by Ingold (1969). *Homolysis* involves the fission of a covalent bond into fragments in which each part carries a single electron, indicated by a heavy dot in (1.3).

$$A\text{---}B \longrightarrow A\cdot + B\cdot \qquad (1.3)$$

Its reverse is *colligation*.

Heterolysis involves the corresponding fission to give fragments, one of which bears both the electrons of the original electron-pair bond. This fragment must, therefore, have become formally one unit more negative, and by its nature be a *nucleophile*, with the power to donate its electron-pair for reaction with an electrophile. The other fragment is formally one unit more positive than it was in the original molecule, and must be electrophilic in character: (1.4).

$$A\text{---}B \longrightarrow A^+ + B^- \qquad (1.4)$$

The reverse of heterolysis is *co-ordination*. Some authors use the term 'addition' in the same sense; but confusion then arises when the reactions of (1.1) and (1.5) both become described as additions.

$$CH_2{:}CH_2 + Cl^+ \longrightarrow Cl.CH_2.CH_2^+ \qquad (1.5)$$

In generalised formulae throughout the remainder of this book, X represents halogen; E^+, an electrophile; Nu, a nucleophile; R, an unspecified substituent (often alkyl or aryl); SOH, a hydroxylic solvent.

1.2 Energy diagrams

In discussing organic reactions one is often concerned with interconversions of quite complex molecules, and with the formation of a mixture of products from a reactant or a set of reactants. Of these products, perhaps only one or two are the desired result of the attempted synthesis or degradation. Most such reactions involve the breaking of covalent bonds and the formation of other bonds. They are therefore *activated processes*; that is, they require *energy of activation* to effect them. Almost any complex organic molecule has a number of isomers; and any of these isomers could theoretically, given enough energy of activation and a suitable reaction path, be transformed one into any of the others. Such an interconversion can be described by an energy (strictly an enthalpy) diagram (fig. 1.1) involving an *initial state* (the reactant), a *transition state*, and a *final state* (the product). Occasionally, an organic reaction can be described in the simplified terms of such a diagram; the energy needed to overcome the barrier between reactants and products is supplied thermally by collision, or in other ways, and the interconversion is a smooth, *synchronous* or concerted reorganisation of the positions of the atoms and bonding electrons with no intermediate energy-minimum between reactant and product. Much more frequently, however, we encounter complicated reaction paths, involving several intermediate

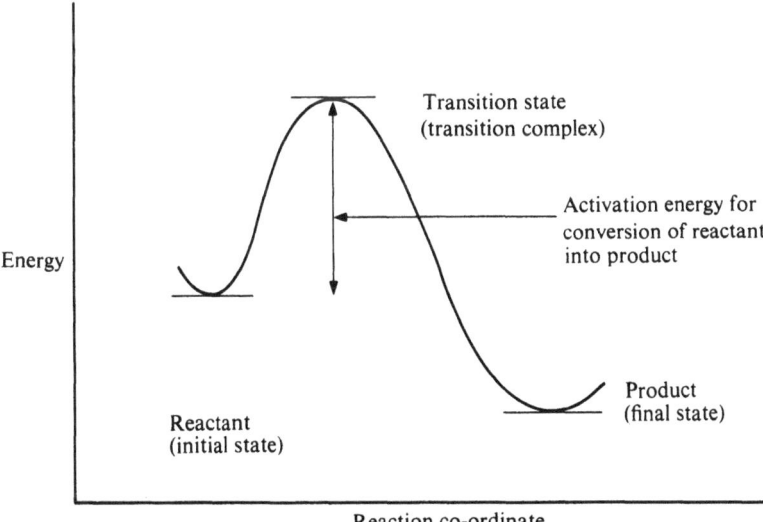

Fig. 1.1. Simplified energy diagram for interconversion of isomers through a pathway not involving an intermediate.

The scale along the reaction co-ordinate may sometimes be approximated by a single physical quantity, such as a changing interatomic distance. Strictly speaking, enthalpies rather than free energies should be represented; see text, pp. 14–15.

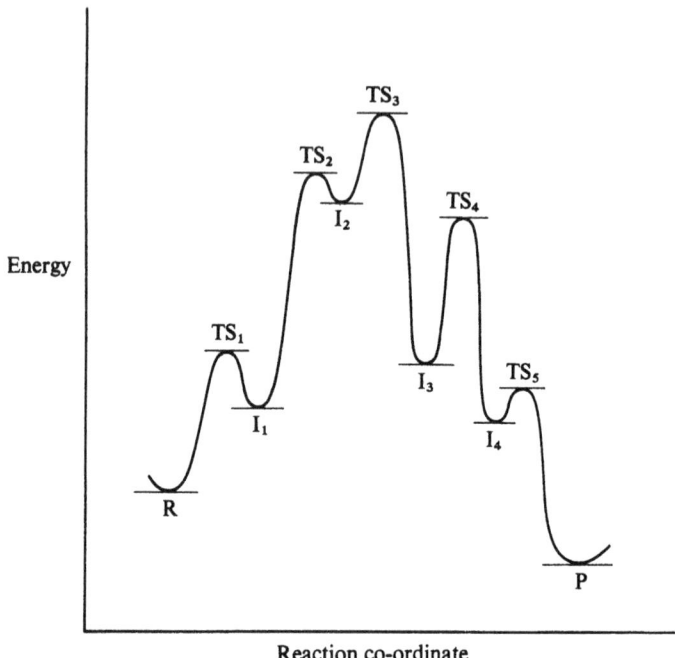

Fig. 1.2. Simplified energy diagram for a reaction involving several intermediates; the scale along the reaction co-ordinate cannot be identified with a single physical quantity. (See also caption to fig. 1.1.)

    R = reactants; P = products; I_1, I_2, I_3, I_4 = intermediates; TS_1, TS_2, TS_3, TS_4, TS_5 = transition states.

stages which we need to understand if we are to control the course of the reaction. Diagrams such as that shown in fig. 1.2 are often used to discuss such sequences, which are typical of the situations that we shall encounter in a study of halogenations. Fig. 1.2 includes several *intermediates* (I_1, I_2 etc.) and several transition states, (TS_1, TS_2, etc.) the transition state of highest energy (TS_3) being that of the *rate-determining* or *rate-limiting* process. That some of these intermediates and transition states may be isomeric is hardly surprising, since isomerism of stable molecules is so common. An approach in terms of such a diagram can be helpful in enabling us to visualise the stages through which the reacting system has to pass, and the relative energetics of the various stages. It should be remembered, however, that such a description is almost always a considerable over-simplification of the proper full physical description of what happens during the reaction. More specialised texts (e.g. Leffler,

1956) should be consulted if the underlying assumptions are to be understood fully; and it is important not to read more into such a diagram than is actually implied by it.

1.3   Intermediates

One of the most difficult problems associated with such descriptions is concerned with the nature and significance of intermediates along the reaction path. It is a basic principle of transition-state theory that the transition state is in equilibrium with the reactants, being converted into products through its instability in the *reaction co-ordinate*, which represents only one of its many modes of change in the relative atomic positions. For some purposes, then, it can be held that any intermediates lying on the left of the rate-determining transition state (fig. 1.2), being themselves in equilibrium with starting materials, are irrelevant to the formation of the transition state. In some complex reactions, however, this view is not tenable, since an equilibrium pre-association may provide the only way in which all the components of the transition state can be brought together.

Not all intermediates holding reactants in pre-equilibrium will necessarily be relevant to the reaction, but to decide which are significant, and in what way they function, is not always straightforward. Fig. 1.3 provides an example. Here we postulate three isomeric positively charged intermediates, and three isomeric products. It may at first sight appear intuitively obvious that the intermediate (**1.1**) leads to product (**1.4**), and that this reaction path either is irrelevant to the formation of the other products or is a stage of diversion which may temporarily form a repository of the reactants before allowing the formation of the isomers (**1.5**) and (**1.6**), thermodynamically more stable than (**1.4**), through reversal of the stages of formation of (**1.4**). This conclusion will be wrong, however, if there exists a route or routes (shown by dashed lines in the figure) whereby (**1.2**) or (**1.5**) can be formed directly from (**1.1**) or (**1.4**). In such a case (as, for example, when $E = NO_2$ (Ingold, 1969; Hughes and Ingold, 1952)), the formation of the intermediate (**1.1**) may be especially relevant to the formation of (**1.5**), and may allow or force the bypassing of (**1.6**) which would normally be preferred over (**1.5**).

A third problem is concerned with methods of defining the intermediates lying to the right of the rate-determining transition state. Strict theory requires that a set of reactants which attain the energy of the transition state pass over the energy barrier because they are unstable only in one single vibrational co-ordinate (the reaction co-ordinate),

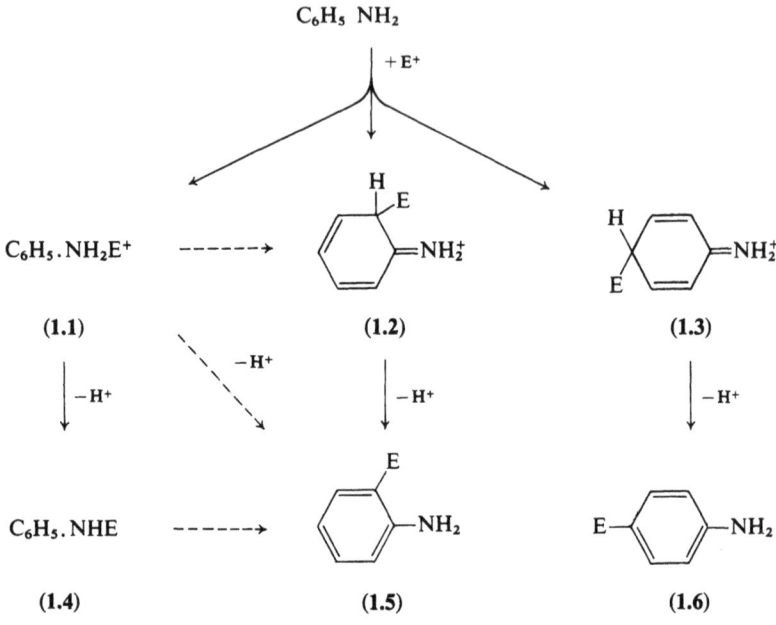

Fig. 1.3. Intermediates in a hypothetical aromatic substitution.

and so proceed onward to products to give only one mode of reaction (Leffler and Grunwald, 1963). In practice, however, some transition states are probably best thought of as a family of closely related structures, able to lose energy to the environment in more than one way, and so capable of more than one mode of decomposition. Furthermore, any intermediate to which the transition state leads may be able to react in more than one way, to give more than one product.

1.4 Methods of mechanistic study

The study of the halogenations which we shall discuss in this book can involve detailed investigations involving a wide range of physical and chemical techniques. These have of recent years been greatly extended in scope and in power through the availability of new spectrometric methods of analysis, together with devices which allow the rapid scanning of changes occurring in a reaction mixture. It will be appreciated that, whereas important examples of some reactions have been studied in very great detail, others have had only cursory or incomplete attention so far. It should be noted also that it is characteristic of many organic reactions that each new investigation produces new information, thus enabling, or requiring, that a relatively simple interpretation be replaced by a more elaborate one.

In beginning the mechanistic study of a reaction, it goes without saying that the structures of the starting materials need to be established, and that any equilibria involving them, or any of them and any component of the environment, should if possible be elucidated. Spectroscopy is often used for studies of this kind. It is then necessary to know the nature of the products, and the stoicheiometry of the reaction; that is, the proportions in which the various components are obtained. Frequently this aspect of the investigation is best associated with kinetic measurements, in which the rate of disappearance of one or more of the reactants is measured along with the rate of formation of one or more of the products. If these are the same, then a relatively simple description of the reaction may be possible. Discrepancies, if they are noted, may provide important information concerning the intervention of intermediates along the reaction path. When intermediates are suspected, they can often be detected by physical methods; as for example when calorimetric or spectrometric procedures are used to follow the course of reaction. Chemical procedures are often used also, as when intermediates can be trapped and diverted to new, recognisable products which help to define the nature of the intermediate.

1.5 Kinetic methods

Kinetic methods help also in allowing deductions to be made concerning the nature of the transition state. This state, through which the reactants must normally pass in order to become the products, cannot be examined directly by the physical methods available for study of intermediates because its lifetime is no greater than that of a molecular vibration or collision. Study of mechanism, however, requires that inferences be drawn concerning its nature. The stoicheiometry of the transition state is normally deduced from the kinetic form of the reaction. A familiar example comes through the chemistry of the alkyl halides, R.X. The reaction of (1.6) can follow the kinetic form of (1.7), in which case it would be concluded that the transition state had the composition [R.X] and did not include the nucleophile Nu$^-$; its structure might then be assigned as in formula (**1.7**), the dots indicating a partially broken R–X bond. Alternatively, the second-order kinetic form, (1.8), may be observed; in which case the transition state must have the composition [R.X, Nu$^-$], and may be assigned the structure shown in formula (**1.8**),

$$R.X + Nu^- \longrightarrow R.Nu + X^- \qquad (1.6)$$

$$d[X^-]/dt = d[R.Nu]/dt = k_1[R.X] \qquad (1.7)$$

$$\overset{\delta+}{R} \ldots \overset{\delta-}{X} \qquad [Nu \ldots R \ldots X]^-$$

$$\textbf{(1.7)} \qquad\qquad\qquad \textbf{(1.8)}$$

$$d[X^-]/dt = d[R.Nu]/dt = k_2[R.X][Nu^-] \qquad (1.8)$$

with a partially formed bond from R to X, and negative charge shared largely between the centres X and Nu.

Powerful as the kinetic method is, yet it must be used with caution, particularly for reactions in solution. Deductions from it depend on being able to vary the concentrations of the reactants and observe changes in rate which are used to identify the composition of the transition state. If, however, this variation in concentration modifies the properties of the medium in which the reaction is occurring, a superimposed influence on the rate may be observed, and this may obscure the true kinetic form.

1.6 Environmental influences; solvent effects

Most organic reactions are carried out in the liquid state. When the solvent is not one of the reactants it may play the vital, but essentially merely mechanical role, of allowing the reactants to come rapidly into the necessary physical proximity; but in doing so it may interact with any or all of them, and it may also interact with any intermediate states between reactant and product, including the rate-determining transition state. Hughes and Ingold (1935) developed a theory of solvent action which covers also the environmental influences of added electrolytes on the basis that if a change in solvent, or other non-reacting components of the environment, stabilises the transition state more than the initial state, then the reaction will be facilitated. This theory is of wide generality; in practice it is not always easy to apply, since reactants and solvents can interact in very specific ways which are difficult to predict.

Organic chemists have a wide range of solvents available to them, including the various hydroxylic solvents (water, the alcohols, phenols, carboxylic acids, etc.), dipolar aprotic solvents (acetone, dimethyl sulphoxide, etc.), and non-polar solvents (benzene, hexane, etc.). Some of the art of organic chemistry lies in the choice of the most appropriate solvent to effect a desired transformation.

When the solvent itself is one of the reactants this introduces special difficulty into the use of the kinetic method for mechanistic deductions, since it then becomes difficult to distinguish between the function of the solvent as a reagent and its function in providing the environment for reaction (Ingold, 1969; Streitwieser, 1962).

1.7   Catalysts

The effects of particular *catalysts* are often helpful in elucidating what mechanisms are available for a reaction. Organic peroxides and other initiators of homolytic processes are often very effective catalysts; they start chain reactions involving free-radicals, and one act of initiation may result in the conversion of many molecules of reactants into products through many repetitions of the chain-propagating steps. This type of catalyst may ultimately be consumed in side reactions. Study of the products, and of the effects of possible inhibitors of the radical-chains, is often necessary to establish whether or not the uncatalysed reaction is following the same pathway as the catalysed process.

In this book we shall be concerned more often with catalysts for hetero-lytic processes. Acids and bases are the most important of these; their modes of functioning are reviewed by Bell (1973). Among acid catalysts we must consider the hydroxonium ion, H_3O^+; other positively charged molecules derived by protonation of the solvent (e.g. $CH_3CO_2H_2^+$, the conjugate acid obtained by protonation of the solvent acetic acid); neutral species which can act as *general acids* (e.g. the acetic acid mol-ecule, CH_3CO_2H); and *Lewis acids* (e.g. $AlCl_3$). For all these types, the catalytic function generally involves the transference of an electrophile reversibly to one of the reactants. In this way is produced a more reactive species which can undergo heterolysis to give a more reactive intermedi-ate; the latter then participates in further transformations. Equations (1.9) and (1.10) illustrate how a proton-donor functions in this way; (1.11) illustrates the familiar, and chemically similar, function of a Lewis acid.

$$\text{t-Bu.OH} + \text{H}^+ \rightleftharpoons \text{t-Bu.OH}_2^+ \qquad (1.9)$$

$$\text{t-Bu.OH}_2^+ \longrightarrow \text{t-Bu}^+ + \text{OH}_2 \qquad (1.10)$$

$$\text{CH}_3.\text{Cl} + \text{AlCl}_3 \rightleftharpoons [\text{CH}_3^+ \quad \text{AlCl}_4^-] \qquad (1.11)$$

Catalysis by bases in organic chemistry usually involves the converse function: namely, the removal of a proton, (1.12), or of another electro-phile, (1.13), to form an intermediate which is more reactive than the starting material.

$$\text{CH}_2(\text{CO}_2\text{Et})_2 + \text{OEt}^- \longrightarrow [\text{CH}(\text{CO}_2\text{Et})_2]^- + \text{HOEt} \qquad (1.12)$$

$$\text{CH}_3.\text{CO.NH}_2\text{Br}^+ + \text{Cl}^- \longrightarrow \text{CH}_3.\text{CO.NH}_2 + \text{BrCl} \qquad (1.13)$$

1.8   Acidity-functions

Acidity-functions have been used for some mechanistic deductions; recent reviews are given by Hammett (1970) and by Rochester (1970). Their use involves an attempt to study acid-catalysed reactions over a range of concentration sufficient to allow disentanglement of the separate influences of changing environment, of departures from ideal behaviour of electrolytes, and of the way in which the solvent is taking part in the reaction. The simplest acidity-function is the stoicheiometric concentration of hydrogen ions; but this measures the power of the medium to donate a proton to a substrate only in very dilute aqueous solution. At high concentrations of mineral acids in water significant deviations can be observed when the protonating power is measured by spectrometric measurement of the concentrations of base and conjugate acid. Typically, these deviations become important when the concentration of mineral acid exceeds about 0.5 M; they vary with the acid, with the solvent, and to some extent with the type of base used for the measurement.

The consequent definition of a useful acidity-function, and its application to a mechanistic discussion, rests on an intuitive assumption that the protonation of the organic substrate concerned in the reaction under mechanistic discussion will be paralleled by the chosen acidity-function. Certainly there are often circumstances in which correlations of this kind will lead to useful deductions, but recent research has not borne out the original hope that acidity-functions would provide information which would enable unambiguous assignment of the role of the solvent in acid-catalysed reactions generally.

1.9   Isotope effects

Potentially a more powerful approach to a study of reaction pathways involves the study of *isotope effects* on equilibria and on reactivity (Melander, 1960). Isotopes differ in their chemical properties only by virtue of their different masses, and so provide a sensitive probe of the vibrational changes experienced by an assembly of molecules undergoing reaction. In principle, a study of the effect of changing one isotope for another in a reactant gives information concerning the extent to which the bond involving the changed atom is modified in going from the initial to the transition state. The biggest differences in masses are found with the isotopes of hydrogen, so hydrogen–deuterium or hydrogen–tritium isotope effects are relatively large and can introduce rate differences of the order of a factor of ten. Heavy-atom isotope effects are much smaller, but their study, now that the appropriate experimental techniques are becoming available, is providing significant information that cannot be obtained any other way (Fry, 1972).

1.10   Structural effects

One of the most important approaches to the nature of organic reaction pathways involves a study of the effects of systematic change in structure of one or other of the reactants on the rate of the reaction. Historically, the use of this method derived from empirical observations that such changes can often be correlated with similar effects observed in other reactions, or with related effects on measurable physical properties such as the dipole moment of the molecule. From these observations sprang the electronic theory of organic chemistry. Ingold was one of its greatest exponents and his terminology has been very widely used, though it has become somewhat modified through the years. He recognised that it was justifiable to a first approximation to separate two types of effect. The first, essentially electrostatic in character and exerted by poles or by dipoles, is transmitted inductively either through space or through the bonding system; the second, essentially involving delocalisation of bonding electrons by quantum-mechanical resonance, is most important for conjugated systems involving multiple bonds and lone-pairs of electrons. He recognised also that these effects could be brought into play differentially, depending on the detailed properties of the reagents and the environment; so he distinguished between permanent effects of polarisation (the inductive and mesomeric effects) and effects of polarisability (the inductomeric and electromeric effects).

In this book, the terms inductive effect ($\pm I$) and conjugative effect ($\pm K$) will be used in their usual senses (Ingold, 1969), but will be interpreted as including the corresponding effects of polarisability which will be discussed separately only where this is necessary. The signs are used conventionally to represent the direction of the effect; electron-releasing groups are assigned a positive effect, and electron-withdrawing groups a negative effect. Formulae (**1.9**) and (**1.10**) illustrate these matters.

$$\overset{\longleftarrow}{\underset{\text{Cl}-\text{CH}_2-\text{CH}_2-\text{CH}_3}{\scriptstyle\delta-\quad\ \delta+\qquad \delta\delta+\qquad \delta\delta\delta+}}$$

The Cl substituent exerts a ($-I$) effect.

(**1.9**)

$$\text{HO}\overset{\curvearrowright}{-}\text{CH}\overset{\curvearrowright}{=}\text{CH}_2$$

Otherwise in valence-bond terms:

$$[\text{HO}-\text{CH}=\text{CH}_2 \longleftrightarrow \text{HO}^+=\text{CH}-\text{CH}_2^-]$$

The HO substituent exerts a ($+K$) effect.

(**1.10**)

We shall need to refer also to *hyperconjugation* (Baker, 1952) normally using this term in its simplest sense, in which it is considered that the electrons of a single bond become partly delocalised into an unsaturated system: formula (1.11). The H_3C substituent exerts a $(+K)$ effect by hyperconjugation; its accompanying $(+I)$ effect is for simplicity not represented in the formula.

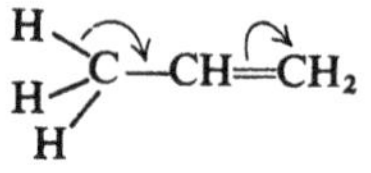

$$\begin{array}{c} H \\ H - C - CH = CH_2 \\ H \end{array}$$

Otherwise in valence-bond terms:

$$\begin{array}{c} H \\ H - C - CH = CH_2 \\ H \end{array} \longleftrightarrow \begin{array}{c} H^+ \\ H - C = CH - \bar{C}H_2 \\ H \end{array}$$

(1.11)

1.11 Linear free-energy relationships

Of recent years there have been many attempts to put the qualitative electronic theory of organic chemistry into more quantitative terms by the use of *linear free-energy relationships*. In this approach, the logarithms of the relative rate coefficients for one reaction of a series of compounds (e.g. $\log(k_R/k_H)$ for the hydrolysis of $R.C_6H_4.CMe_2.Cl$, R being varied) are plotted against the corresponding values for another reaction (e.g. the reaction of $R.C_6H_4.H$ with chlorine). A straight-line relationship indicates that the substituent R is affecting the rates of the two reactions in related ways: the slope of the line gives a comparison of the relative response of the two reactions to change in structure; and the sign of the slope tells whether or not a particular structural change favours both reactions. Hammett (1938) used the dissociation constants, K_a^R, of substituted benzoic acids, $R.C_6H_4.CO_2H$, to define a series of *substituent constants*, σ_R, through the relationship of (1.14).

$$\sigma_R = \log_{10}(K_a^R/K_a^H) \qquad (1.14)$$

From this is derived the *Hammett equation*, (1.15).

$$\log_{10}(k_R/k_H) = \rho\sigma_R \qquad (1.15)$$

For a reaction for which this treatment gives a satisfactory linear plot the influence of substituents on the rate can be satisfactorily dissected into an influence dependent only on the substituent (measured by the substituent constant σ_R), and an influence dependent only on the reaction (measured by the reaction constant ρ). Because of Hammett's choice of standard reaction, positive values of the substituent constant are to be associated with electron-withdrawing substituents, and negative values with electron-releasing substituents. Values of ρ are greater than 1 when the rate of reaction responds to change in the substituent more than in the standard reaction, and are positive if the change is in the same direction, namely that electron-withdrawing substituents facilitate the reaction.

This type of treatment gives reasonable correlations over a modest range of reactions; but it turns out not to be applicable very widely, except to a rather poor approximation. This might have been expected, since we are attempting to describe the effect of any single substituent with only a single constant, whereas the qualitative theory found it necessary to use four different effects in its descriptions. There have been a number of attempted refinements, reviewed by Ingold (1969); and different sets of substituent constants, better applicable to particular situations, have been tabulated. One of the most useful of these sets in the context of this book involves the use of (1.16), in which the standard reactions chosen, with rate coefficients under specified conditions k_R^{st}, are the unimolecular solvolyses of substituted aryldimethylcarbinyl chlorides, $R . C_6H_4 . CMe_2Cl$.

$$\sigma_R^+ = -4.52 \ \log_{10}[k_R^{st}/k_H^{st}] \qquad (1.16)$$

Here the negative sign and the numerical constant (-4.52) are included in one definition in order that the numerical values of σ_R^+ should resemble those of σ_R (see (1.14)) for substituents having minimal conjugative effects. Substituent constants, σ_R^+, defined in this way correlate reasonably well with the rates of a number of electrophilic halogenations (Stock and Brown, 1963).

1.12 Stereochemical influences; steric effects

The term *stereochemistry* is almost as wide-ranging as chemistry itself. Applied to structure, it includes everything that determines the geometry of a molecule, including bonding forces and electronic and entropic factors, varying with temperature according to the laws of statistical mechanics, which determine and may partly restrict the flexibility of the molecule. Applied to a reaction, it refers to the detailed geometry of reactants, intermediates, transition states and products. Stereochemistry therefore provides an important guide to the mechanism of a reaction.

The electronic (or 'polar') effects of organic chemistry (i.e. the inductive and conjugative effects already referred to) are often usefully distinguished from steric effects, and both can be of great significance in determining the ease of a reaction. It is customary to distinguish between *primary steric effects*, which result from direct steric hindrance or acceleration of the reaction and arise because the initial and transition states differ in their degrees of congestion; *secondary steric effects*, which arise because a substituent restricts intramolecularly the geometry of another substituent, the geometry of which determines its ability to exert a polar (usually a conjugative) effect; and *stereo-electronic effects*, which occur when the movement of bonding electrons on going from initial to transition state has special geometric requirements.

1.13   Effects of temperature

Most reactions proceed faster when the temperature is raised, as would be expected for a process requiring thermal excitation. When a one-stage reaction is under observation (fig. 1.1) important information can be obtained by studying the rate of reaction as a function of temperature. Then by using the Arrhenius equation, (1.17), two separate physical quantities can be evaluated. One is the *Arrhenius energy of activation, E,*

$$\log_{10} k = Be^{-E/RT} \qquad (1.17)$$

which can be identified approximately with the energy required to activate the initial state to the transition state. The other is the quantity B; for many reactions this, like E, is experimentally nearly independent of temperature. It can be regarded as a measure of the probability that the reactants, having been activated by the energy of activation, will actually pass over the energy barrier and form the products. Alternatively, if one prefers to avoid the conceptualisation implicit in the collision theory of reaction, it can be regarded as a measure of the *entropy of activation* for the reaction.

When a multi-stage reaction (fig. 1.2) is being considered, however, interpretation of the change in rate with change in temperature is more complicated. For example, the effect of temperature on the various pre-equilibria involving the reactants may be so profound that the whole reaction may show a negative temperature coefficient; a negative activation energy for an activated process is, of course, physically an invalid concept. This is one of the reasons why many mechanistic comparisons based on interpretation of structural effects are made at a single temperature: it often seems useful to assume that the influences of structural change can be seen more clearly in relative rates (i.e. in free energies

of activation) than in the relative energies and entropies of activation
studied separately. It is also one of the reasons why, in diagrams such as
fig. 1.1 and fig. 1.2, enthalpies or free energies of activation are often
presented rather indiscriminately and without remarking on possible
differences between them.

**1.14   Thermodynamic and kinetic factors as determinants of rates and
products**

For a one-stage reaction the transition state, which separates reactants
from products, must have some of the properties of both. It is easy to fall
into a misconceived way of thinking based on the rather natural feeling
that the greater the thermodynamic stability of the products relative to
the starting material, the easier will be the reaction. In general, this is not
the case. Thus many thermodynamically unfavourable processes (e.g. the
protonation of many oxygen bases) can be very rapid; conversely, many
thermodynamically favourable processes (e.g. the reaction of molecular
hydrogen with molecular oxygen) can be very slow. On the other hand,
other things being equal, a change in a reaction system towards making
the products thermodynamically more stable relative to the reactants
should make the reaction more rapid. In fig. 1.4, the reaction co-ordinate

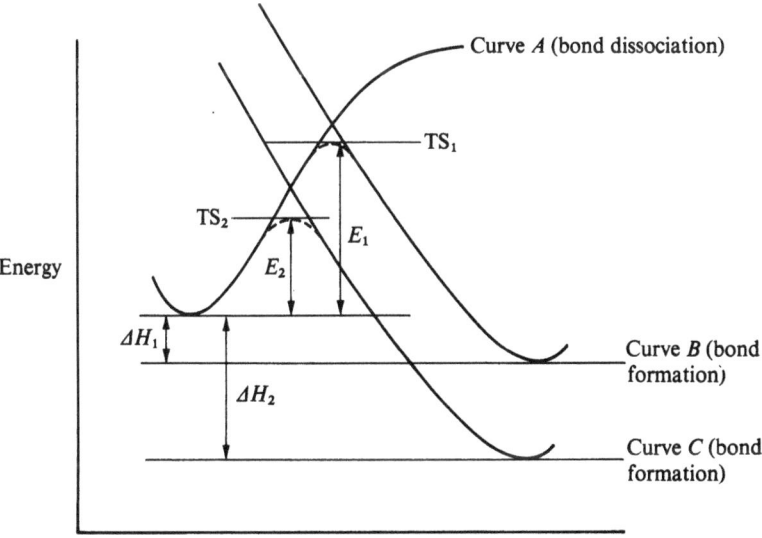

Fig. 1.4. Relationship between heat of reaction (ΔH), energy of activation (E) and
position of transition state (TS) on the reaction co-ordinate.

(shown dotted) is approximated by a transition from bond breaking (curve A) to bond formation (curves B, C). The course of the reaction then is changed by changing the energy of bond formation. When this is increased (as in the change from B to C), the heat of reaction increases ($\Delta H_2 > \Delta H_1$) and in consequence the energy of activation decreases ($E_2 < E_1$). At the same time the transition state TS_2 is moved on the reaction co-ordinate in such a direction that it becomes more reactant-like than TS_1, a result which is not intuitively obvious. This type of approach was discussed by Evans and Polanyi (1938), and more recently by Hammond (1955); it can be used also to compare the selectivity of reagents. Deductions made on these lines can be relied on only when things really are equal, as for isotopic substitution of a bond formed or broken in the reaction or for other relatively small changes in one atom or bond concerned in a reaction.

When an organic reaction gives more than one product, the products are usually not formed in thermodynamic equilibrium: often (though not always), much more than the equilibrium amount of the thermodynamically unfavoured product is obtained. Aromatic substitutions provide a familiar example: the *meta*-substituted product is often the thermodynamically most stable of the isomers formed, but the *ortho*- and *para*-derivatives often predominate in the products. We then say that the products are formed under *kinetic control*. When equilibrium is reached (which in the cited case is often difficult to realise, but in others is relatively easy), the products are said to have been formed under *thermodynamic control*.

1.15 Classification of mechanisms

There is no agreed and perfectly satisfactory classification of mechanisms; individual authors have often adopted their own symbolisms. We have already noted some of the important divisions into classes. In this book we shall not be much concerned with heterogeneous reactions (that is, those occurring at the interface between phases), nor with reactions in the gas or in the solid phase; but instead with *homogeneous* reactions in solution, and in particular with heterolytic electrophilic halogenation: that is, reactions in which the halogenating reagent acts as an electrophile.

The molecularity of individual stages of a reaction can often be used for classification: a reaction is *unimolecular*, for example, if only one molecule is necessarily undergoing covalency change in the rate-determining transition state. Ambiguities and blurred distinctions arise in using this type of classification, particularly in two types of situation. The first is when there are strong interactions between the solvent and either the initial or the transition state. The second is when this method of classification is used for the rate-determining stage of a multi-stage reaction. Then it may happen that two stages of a reaction are so close in rate that they become jointly rate determining, or it may not be clear how many particles should be considered as comprising the transition state.

Classification in terms of *kinetic order* is best regarded as a classification of an experimental, rather than of a fundamental theoretical kind. Thus a solvolytic process of the first kinetic order may involve one, or two, or more particles in the rate-determining step, and so the kinetic order itself indicates only part of what is happening in the transition step.

It often turns out to be valuable to distinguish between *intermolecular* and *intramolecular* processes. In the former, the reaction involves covalency changes in two separate molecules; in the latter, two or more reaction sites within the same molecule are concerned.

The time sequence of events is also mechanistically important. In some reactions, bond-making and bond-breaking processes are *synchronous* or *concerted*. Here the energy of forming new bonds partly offsets the energy of breaking old ones. These processes, of course, proceed in such a way as to minimise the energy of the transition state; so bond forming and bond breaking need not have proceeded to the same extent, and change in structure may produce a different balance between them. In other reactions, the individual stages are discrete, with recognisable intermediates intervening. It then becomes necessary to specify not only that the reaction is stepwise, but also, if possible, the order in which the stages occur. To make the latter distinction may often be experimentally difficult.

1.16 Additions, substitutions, replacements, displacements and halogenations

In this book, the term *addition* will be used to imply a reaction in which *two* new covalent bonds are formed to a substrate. Most of the reactions discussed in connection with the reactions of the halogens (e.g. (1.1)) are 1,2-additions in which both ends of a double bond become saturated, or a triple bond becomes converted into a double bond. Other modes of addition (e.g. 1,1- and 1,3-addition) are possible, however, and it should be recognised that not all authors use the same scheme of classification. Additions are sometimes accompanied by double-bond rearrangement; the 1,4- and 1,6-additions to conjugated systems are of this kind.

It is thought by some writers that the term 'substitution' should be restricted to reactions in which a hydrogen is replaced by a substituent. This usage has, however, become generalised by many workers, particularly in the context of reactions in which one nucleophile is replaced by another. The terms *substitution*, *replacement* and *displacement* will, therefore, be used without any necessary implication concerning the nature of the leaving group.

When the term *halogenation* is used, no distinction will be made as to whether a co-ordination, a replacement or an addition is implied; and halogen will be thought of as the reagent, not as the substrate. So an *electrophilic halogenation* is one in which, in the sequence of events which leads to reaction, the first process in which a source of halogen attacks the substrate involves electrophilic halogen.

1.17 Symbolism

Conventional formulations are used for symbolism in this and subsequent chapters; some of the generalisations are mentioned in §1.1. Structures **(1.10)** and **(1.11)** show two ways in which resonance hybrids are represented; each implies electron delocalisation resulting in stabilisation. A third, which is sometimes convenient because it is concise, is to represent partial bonds as in **(1.12)**. Aromatic molecules are usually represented by using a single canonical structure, but it is sometimes helpful to indicate cyclic resonance as in **(1.13)** or **(1.14)**. Dotted lines are also used to represent bonds partly formed or partly broken in a transition state, as in **(1.7)** and **(1.8)**. Dashed lines and wedges are used to represent stereochemistry; in **(1.15)**, the full lines represent bonds lying in the plane of the paper; the group X lies behind, and the group Y in front, of that plane.

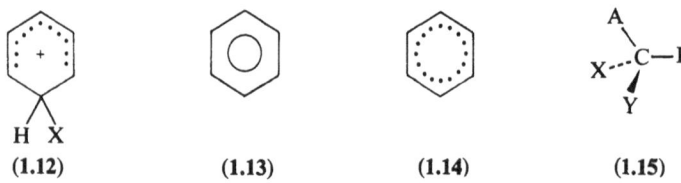

(1.12)	**(1.13)**	**(1.14)**	**(1.15)**

2 The bonding and non-bonding properties of the halogens

'This imperfect sketch...concerning the structure...'
(Oxford English Dictionary, vol. x, p. 1165)

2.1 Introduction

The halogens comprise group VII of the periodic classification of the elements. Each of the atoms has one electron fewer than in a noble gas; the electronic configuration of this gas can, therefore, be achieved by gain of a single electron to form a uninegative ion, or by the sharing of an electron to form a single covalent bond. Some of the properties of the halogens and of their immediately derived forms are given for reference in table 2.1; a recent comprehensive account by Downs and Adams (1973) documents much of the factual material in this chapter.

Monotonic series with increasing atomic number are to be noted for some of the properties listed in table 2.1: for example, the covalent radius, the van der Waals radius, the ionisation enthalpy and the bond enthalpy of the carbon–halogen bond are all properties which change regularly down the series. Other properties, however, show irregularities in which fluorine is notably anomalous. Such anomalies probably reflect the irregular increments in nuclear charge (8, from F to Cl; 18 from Cl to Br; 18 from Br to I; 32 from I to At) together with irregular changes in the shielding of the effects of this charge on the non-bonding and bonding shells of electrons. Thus the intrinsic energy gain involved in allowing the element fluorine to attain the configuration of an inert gas is partly offset by electrostatic repulsions if the extra electron has to be accommodated too close to the atomic nucleus, and so the electron affinity of fluorine and the strength of the F–F bond in the fluorine molecule are both less than the corresponding values for chlorine.

Fluorine, chlorine and bromine would normally be considered as essentially non-metallic in character; iodine is the first of the halogens which would be thought of as potentially metallic. For the process shown

19

TABLE 2.1 *Some properties of halogen atoms, molecules and ions*[a]

Element (X)	Atomic number	Adjacent noble gas	Electronic structure	Electron affinity (298 K) /mol kJ^{-1}	Electro-negativity (Allred-Rochow scale)	Ionisation enthalpy ΔH_{298}/kJ mol^{-1}	Covalent radius /pm[b]	Van der Waals radius/pm[c]	b.p. (X$_2$)/°C	m.p. (X$_2$)/°C	Bond enthalpy (X$_2$) ΔH/kJ mol^{-1}	Bond enthalpy (C-X) ΔH/kJ mol^{-1}	Hydration enthalpy (X$^-$) ΔH/kJ mol^{-1}
Fluorine (F)	9	Ne	$1s^22s^22p^5$	339	4.1	1682	71	131	−188	−223	159	485	506
Chlorine (Cl)	17	Ar	$[Ne]3s^23p^5$	355	2.8	1255	99	181	−34.6	−103	243	331	368
Bromine (Br)	35	Kr	$[Ar]3d^{10}4s^24p^5$	331	2.7	1142	114	196	58.8	−7.2	192	276	335
Iodine (I)	53	Xe	$[Kr]4d^{10}5s^25p^5$	301	2.2	1008	133	222	184.4	113.5	151	238	293
Astatine (At)[d]	85	Rn	$[Xe]4f^{14}5d^{10}6s^26p^5$	(ca 297)	2.0	(ca 920)	(ca 140)	(ca 230)	Not known	Not known	(ca 127)	Not known	(ca 276)

[a] Most of the numerical values are from standard tabulations (cf. Downs and Adams, 1973). The conversion factor 1 kcal = 4.184 kJ has been used.
[b] Twice the X–X bond distance.
[c] The ionic radii have nearly the same values.
[d] Values in brackets have been estimated by extrapolation.

in (2.1), equilibrium constants $(K/\mathrm{mol}\,\mathrm{l}^{-1})$ of 10^{-40}, 10^{-30} and 10^{-21} have

$$X_2(\mathrm{aq.}) \rightleftharpoons X^+(\mathrm{aq.}) + X^-(\mathrm{aq.}) \tag{2.1}$$

been estimated where X = Cl, Br and I respectively. It might have been thought, therefore, that by reducing sufficiently the concentration of X^-, the free iodine cation might have become accessible in solution. It seems, however, that as yet no evidence has been adduced to establish the existence of free Cl^+, Br^+ or I^+ as distinct from species (some of which are described in §2.7) in which these halogen cations are stabilised by covalent co-ordination with a nucleophile. Whether the astatine cation is accessible remains unknown.

The stereochemistries associated with compounds of the halogens illustrate the marked contrast between the chemistry of fluorine and those of the other halogens. Fluorine is virtually unable to form compounds having oxidation states greater than zero; it also cannot exceed a co-ordination number of two (one, if hydrogen-bonding is excluded) in covalently bonded systems. The higher halogens, on the other hand, can exist in various oxidation states, and can have co-ordination numbers which involve the accommodation of more than two groups (for iodine, of up to seven groups) covalently bound to the central atom. In table 2.2, some of the important compounds and their stereochemistries are given.

A description of the bonding in these compounds can be given in terms of two-electron covalent bonds. On this model, it is necessary for

TABLE 2.2   *Stereochemistries of some compounds of the halogens*

Co-ordination number of central halogen	Stereochemistry	Examples
1	Diatomic unit	F_2, Cl_2, Br_2, I_2, $BrCl$, HCl
2	Linear	I_3^-, $Cl-I-Cl^-$, $Br-I-Cl^-$
2	Angular	ClO_2, ClO_2^-, BrF_2^+
3	Trigonal pyramid	ClO_3^-, BrO_3^-, IO_3^-
3	T-shaped	ClF_3, BrF_3, $RICl_2$
4	Tetrahedral	ClO_4^-, BrO_4^-, IO_4^-, Cl_2O_7, $FClO_3$
4	Square pyramid	ICl_4^-, I_2Cl_6
4	Trigonal pyramid	$IO_2F_2^-$, IF_4^+
5	Square pyramid	ClF_5, BrF_5, IF_5
6	Octahedral	H_5IO_6, OIF_5
7	Pentagonal bipyramid	IF_7

those compounds containing a halogen having co-ordination number greater than one to assume that the bonding shell of electrons in the central halogen contains more than eight electrons: its octet is thus expanded, and d-orbitals must be considered to be involved in the bonding hybridisation. The observed geometrical patterns of bonding can then be interpreted, though probably only in part predicted (Sidgwick and Powell, 1940; Gillespie and Nyholm, 1958).

An alternative approach (Downs and Adams, 1973; Deb and Coulson, 1971) is to regard the stable octets of the halogens as attained but not exceeded in their compounds. Many of the compounds must then be regarded as having some bonds in which an average of fewer than two electrons per bond contribute to the bonding stabilisation. Either a valence-bond or a multicentred-orbital description can then be used, as in the corresponding descriptions of other more obviously electron-deficient molecules. There are at least two schools of thought on these matters; in this book, the representation of ICl_3 as in structure (**2.1**) implies I–Cl bonds having significant (usually substantial) covalent character, but does not necessarily imply that each bond is associated with a pair of electrons. We return to this problem in §2.5 when dealing with the trihalide ions.

$$Cl - I - Cl$$
$$|$$
$$Cl$$

(**2.1**)

2.2 The elements and derived diatomic molecules

As we proceed from the lighter to the heavier halogens, the characteristic increase in colour of their covalent compounds is believed to arise because of a progressive shift towards longer wavelengths of electronic absorption bands involving the excitement of the molecules to states in which charge transfer is more important. The fluorine molecule, because of its low bond strength, is very reactive. It readily dissociates into atoms, and then the resulting fragments participate in chain reactions which have the result of burning up organic molecules to form hydrogen fluoride and fluorocarbons. The exothermicity of these reactions makes it possible to envisage their use for rocket propulsion. Chlorine and bromine, however, are generally less indiscriminate in their behaviour, and show progressively diminishing reactivity; thus it is easy to activate photochemically the chain reaction between chlorine and such an inert paraffin as neopentane, (2.2), but the corresponding reaction with

$$Cl_2 + C(CH_3)_4 \longrightarrow (CH_3)_3C.CH_2Cl + HCl \qquad (2.2)$$

bromine is difficult. Iodine is still less reactive than bromine; partly because it forms rather weak bonds to other elements, and partly because it is large enough to cause steric congestion in the products of many organic reactions.

Astatine has a number of isotopes which have been identified as short-lived branch products from the decay of uranium and thorium and from artificially induced nuclear reactions of other kinds. The isotope of longest life is ^{210}At, which has a half-life of only 8.3 h. Macroscopic quantities are, therefore, not available for study. Tracer experiments show that the element is, like iodine, volatile and soluble in benzene and in carbon tetrachloride. It can be reduced by sulphur dioxide (but not by the ferrous ion) to At^-, and it can be oxidised to HOAt and thence to AtO_3^-. Little else is known about it, though some remarks concerning possible positive species derived from it are made in §2.7.

2.3 Mixed halogens

The existence of a number of types of interhalogen compounds has already been noted in table 2.2. Many of them, including the diatomic molecules with which we are here mainly concerned, can be obtained by reaction of the parent halogens in stoicheiometric amount, either in the gas phase or in solution. Some of the compounds obtained in this way are relatively stable thermodynamically, and can be isolated in the crystalline state. Chlorine monofluoride is an example of a compound of this type: (2.3). Others are relatively unstable; bromine chloride disproportionates

$$Cl_2 + F_2 \; \xrightleftharpoons{\quad\quad} \; 2ClF \qquad (2.3)$$

relatively easily to give bromine and chlorine, (2.4), whereas bromine monofluoride disproportionates to give bromine and bromine trifluoride, (2.5).

$$Br_2 + Cl_2 \; \rightleftharpoons \; 2BrCl \qquad (2.4)$$

$$Br_2 + BrF_3 \; \xrightleftharpoons{\quad\quad} \; 3BrF \qquad (2.5)$$

The reactions of the diatomic interhalogen compounds are generally similar to those of the halogens, and can often be recognised as distinct from those of their products of dissociation or disproportionation. Iodine monochloride fits into this category, and has an important use in organic analytical chemistry in consequence. A solution of iodine monochloride can be prepared by dissolving iodine in acetic acid and passing chlorine into the resulting solution until a marked change of colour is observed. The resulting solution (Wijs' solution) can be used for estimating the degree of unsaturation of relatively reactive unsaturated compounds; it behaves as if it were reacting as $I^{\delta+}$–$Cl^{\delta-}$, the iodine being the electrophile because it is the less electronegative halogen of the pair. Other diatomic interhalogen compounds behave similarly.

Ionic modes of dissociation and disproportionation are also sometimes possible. Thus the conductivity of molten iodine can be described in terms of the equilibrium given in (2.6). Similar modes of dissociation can

$$3I_2 \rightleftharpoons I_3^+ + I_3^- \qquad (2.6)$$

be envisaged for the higher mixed interhalogen compounds (e.g. (2.7)).

$$2IF_3 \rightleftharpoons IF_2^+ + IF_4^- \qquad (2.7)$$

2.4 Charge-transfer complexes of the diatomic molecules

Solutions of iodine in solvents such as cyclohexane are violet; the corresponding solutions in hydroxylic solvents are brown. Spectral changes of this kind arise when charge-transfer complexes between the halogen and the solvent are formed. These can be formulated as resonance hybrids of the kind shown in structures (2.2) and (2.3). The ground state is a hybrid

$$[X_2 \ S] \longleftrightarrow [X_2^- \dots S^+] \qquad Me_3N \dots I{-}I \qquad Me_3\overset{+}{N}{-}\overline{\overline{I{-}I}}$$
$$(2.2) \qquad\qquad (2.3) \qquad\qquad\quad (2.4) \qquad\qquad (2.5)$$

to which (2.3) makes a relatively small contribution; the excited state is stoicheiometrically isomeric, with (2.3) making a larger contribution. The absorption bands associated with the excitation are often intense, and may lie in or near to the visible region of the spectrum.

Although chlorine and bromine do not show such big colour changes in solution, they also can form charge-transfer complexes with nucleophilic solvents, and a number of such complexes have been isolated in the crystalline state. Usually, 1:1 complexes are formed, and they are often stable only at low temperature. Examples include complexes formed by association with σ-donors, which possess formally non-bonding electrons, as in the examples of amines (structure (2.4)), oxygen bases, sulphides and some halogen compounds; and those involving π-electron donors, where the donor function is exerted by formally bonding electrons, as in the cases of complexes between halogens and aromatic compounds. In the solid state the (donor–X–X) system is usually linear, as is indicated in structure (2.4) for the complex involving trimethylamine and iodine. For the complex between chlorine and benzene, the halogen lies along an axis perpendicular to the centre of the ring, and so appears to be attached to the π-electron system generally, rather than to a specific carbon atom or bond within the aromatic ring. It is generally assumed, though it cannot be considered to be at all certain, that the same geometries persist in the liquid state.

This type of molecular association is energetically rather weak, and does not involve much perturbation of the halogen–halogen bond length, which for example is 228 pm in bromine, and is only 231 pm in the dioxan–bromine complex. It is therefore more common to represent such complexes rather loosely in some such way as in structure (**2.4**), rather than by using the symbol for a covalent bond, as in structure (**2.5**). The two modes of representation do, however, reflect extremes in the implied strength of the covalent bonding, and are not necessarily different in kind.

It should be recognised, too, that for any particular donor–acceptor pair, isomeric possibilities often exist. Thus structures (**2.6**)–(**2.10**) all represent 1,1-complexes between bromine and anisole; only one of these involves anisole acting as a π-donor, and only one of them involves oxygen as the donor centre. In other chapters we shall see that questions relating to such distinctions can have important chemical significance; and even more subtle examples are possible for quite simple cases.

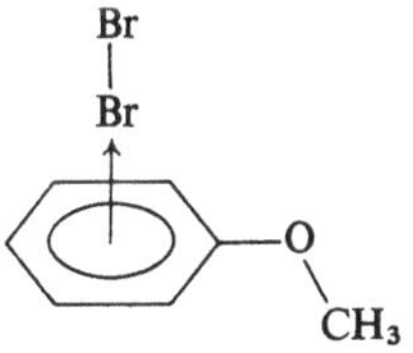

(2.6) 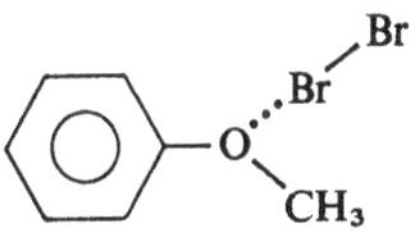   (2.7) 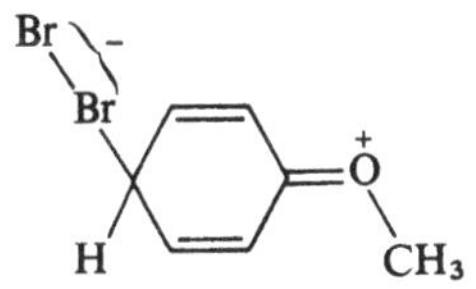   (2.8)

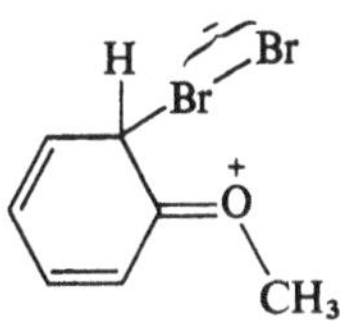

(2.9) 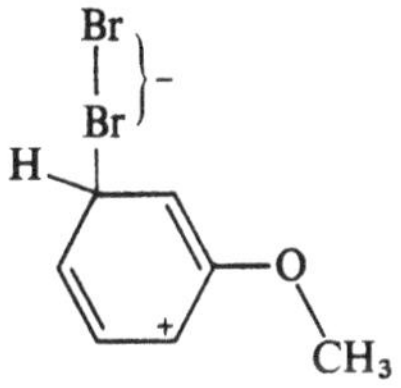   (2.10)

2.5 Polyhalide ions

The 1,1-complexes described above involve charge transfer between an electrophilic halogen molecule and a neutral nucleophile. A similar

TABLE 2.3 *Formation constants (K_f; equation (2.8)) for some trihalide ions at 25 °C in solvents as specified*

Trihalide ion	K_f/mol^{-1} dm^3 (H$_2$O)	K_f/mol^{-1} dm^3 (MeOH)	K_f/mol^{-1} dm^3 (BrCCl$_3$) ·
I$_3^-$	710	—	—
Br$_3^-$	16.3	177	16 600
Cl$_3^-$	0.19	—	—

interaction would be expected between a halogen molecule or an inter-halogen compound and a suitably nucleophilic anion. The most familiar examples comprise the trihalide ions, of which a number of types have been isolated as crystalline solids; many others have been authenticated by the classical methods of physical chemistry. In these compounds, iodine and bromine are known to show a co-ordination number of 6 in the IF$_6^-$ and BrF$_6^-$ ions respectively; chlorine shows a co-ordination number of four in ClF$_4^-$; and no ions having fluorine as the central element are known. Their ease of formation, therefore, increases with the size of the central element in the Hal$_3^-$ combination. Further illustration of this point is given in table 2.3 through values of the equilibrium constants, (2.8), for the reaction of (2.9) and its analogues.

$$K_f = [\text{Br}_3^-]/[\text{Br}_2][\text{Br}^-] \tag{2.8}$$

$$\text{Br}_2 + \text{Br}^- \; \rightleftharpoons \; \text{Br}_3^- \tag{2.9}$$

Study of the trihalide ions by X-ray crystallography, or by infrared and Raman spectroscopy, shows that many of them are linear, or nearly so, with the exterior atoms bonded in an equivalent way to the central atom. Cases exist, as with CsBr$_3$, where this symmetry is distorted, probably as the result of forces associated with crystal-packing. Two possible formulations can be devised for the linear centro-symmetrical trihalide ions. The first (Rundle, 1962) is analogous to the formulation **(2.2)** ↔ **(2.3)** of neutral charge-transfer complexes, and can be illustrated for the tribromide ion by the structures **(2.11)** ↔ **(2.12)**.

$$[\text{Br–Br Br}^-] \quad \longleftrightarrow \quad [\text{Br}^- \text{ Br–Br}]$$
$$\textbf{(2.11)} \qquad\qquad\qquad \textbf{(2.12)}$$

Here the linearity is expected on electrostatic grounds, since the terminal atoms which bear the bulk of the negative charge would prefer to be as far apart as possible. The symmetry derives from the use of two equivalent mesomeric structures, which together provide stabilisation through 'no-bond resonance' (as in the related structures concerned in hyper-conjugation, §1.1, and in electron-deficient molecules generally). Two electrons are used to form two bonds, each of which has a formal bond order of one-half. In terms of orbital hybridisation, the lone pairs on the central halogen atom can be considered to lie in sp^2-hybridised orbitals, leaving a three-centred bonding orbital of p-type to accommodate a further pair of bonding electrons (structure (**2.13**)).

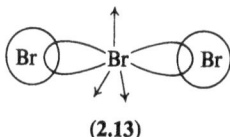

(**2.13**)

This treatment, whether put in valence-bond or in molecular-orbital terms, accords qualitatively with the facts that the stretching frequencies of the X–X bonds in the trihalide anions are lower than those of the X–X bonds in the corresponding diatomic molecules, and that the bond lengths of the symmetrical trihalide ions are significantly greater than those of the halogen molecules (table 2.4). Results obtained by measurements of nuclear quadrupole resonance show also that a very large proportion of the total charge lies on the terminal atoms of the trihalide complex.

It is as well to be aware, however, of an alternative formulation which is to be found in many texts and articles. This derives from generalisations (Sidgwick, 1927; Sidgwick and Powell, 1940) concerning the relationship of electronic structure to the periodic properties of the elements and their compounds. It is assumed that, whilst fluorine cannot easily expand its octet by including d-orbitals in hybridisation with s- and p-orbitals, the higher halogens become increasingly able to do so. In the trihalide ions, therefore, it is presumed that the central halogen accommodates ten electrons by using a set of sp^3d-hybridised orbitals. These theoretically

TABLE 2.4 *Some comparisons of properties of halogen molecules and related anions*

Halogen (X)	Bond length/pm		Principal stretching force constant/nm^{-1}	
	X_2	X_3^-	X_2	X_3^-
Cl	199	—	372	96
Br	228	253	248	91
I	267	290	172	71

might adopt either a trigonal bipyramidal or a square pyramidal arrangement; two of the electron-pairs disposed in them are used for bonding. Several geometries are then possible (e.g. those shown in structures (**2.14**)–(**2.17**)) in which the approximate directions of the axes of the lone-pairs of electrons are shown by arrows). The use of the first of these can be rationalised, if it is assumed that the Br–Br bonds repel each other more than they repel lone-pairs.

$$
\left[\text{Br—Br—Br} \right]^- \quad \left[\text{Br—Br} \atop \text{Br} \right]^- \quad \left[\text{←Br→} \atop \text{Br} \right]^- \quad \left[\text{Br} \atop \text{Br} \text{Br} \right]^-
$$

  (**2.14**)   (**2.15**)   (**2.16**)   (**2.17**)

It is probable that a full description, in which the electronic states of both bonding and non-bonding electrons were included in computing the properties of the systems, would by combining elements from both theories provide a still more accurate approximation.

It should be noted that, whichever of these theories is preferred, the energy differences between angular (e.g. (**2.15**), (**2.16**)) and linear (e.g. (**2.14**)) states might well not be very large. One complex is known to be angular in the solid state. This is the I_5^- anion in Me_4NI_5, which has the geometry shown in (**2.18**). For this complex, because of its size and of the polarisability of its components, electrostatic forces within the ion itself might be expected to be of relatively small importance.

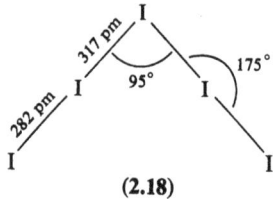

(**2.18**)

2.6   The halide anions: covalent halides as sources of nucleophilic halogen

Compounds having covalent bonds to halogen exist in profusion; in principle, any such compound can react to give a halide anion, either by heterolysis, (2.10), or by replacement, (2.11).

$$R\text{—}X \rightleftharpoons R^+ + X^- \tag{2.10}$$

$$Nu^- + R\text{—}X \rightleftharpoons Nu\text{—}R + X^- \tag{2.11}$$

Both kinds of reaction are in principle reversible, and for any given example the position of equilibrium depends on a number of factors, including the strengths of the covalent bonds in reactants and products and the solvation energies of the ions and molecules concerned in the reversible process. The physical constants summarised in table 2.1 illustrate that the solvation energies of the halide anions in water are large, being similar in magnitude to, and often greater than, the bond energies of the covalent bonds to halogen in the substrates from which they are derived. For this reason, equilibrium positions in reactions involving halide anions are heavily dependent on forces of solvation, and may vary differentially with change in the solvating power of the solvent (Parker, 1962).

The kinetics of processes such as those represented in (2.10) and (2.11) present, as with all chemical reactions, additional problems quite distinct from those involved in determining thermodynamic stabilities and hence positions of equilibrium. Thus at 25 °C in water or in acetone the ionisation, (2.10), of methyl bromide (CH_3Br) is slow, but that of t-butyl bromide (Me_3CBr) is many powers of ten more rapid. Conversely, however, for the corresponding reactions with chloride ions in acetone ((2.11) with Nu = Cl), the situation is the reverse: the reaction of t-butyl bromide is very slow, but that of methyl bromide is many times faster. These comparisons illustrate that relatively small structural changes can have major effects on reactivity, and can operate in opposite directions even when two not too dissimilar reactions are considered. Other works should be consulted concerning the particular comparisons chosen for illustration; here we seek only to remind ourselves that reaction rates depend crucially on mechanistic details.

A further point can be illustrated from reactions of this kind, namely that similar chemistry can arise from reaction involving different charge-types. Thus the reaction of (2.12) with its many analogues in the chemistry of the metallic and non-metallic polyhalides, is similar electronically to that of (2.10). Similarly, (2.13), (2.14) and (2.15) present electronic

$$MCl_4^- \rightleftharpoons MCl_3 + Cl^- \qquad (2.12)$$

analogies with (2.11). The same generalisations concerning thermo-dynamics and kinetics apply in all these cases; the detailed effects of

$$H_3N + R\text{—}X \rightleftharpoons [H_3NR^+] + X^- \qquad (2.13)$$

$$H_2O + MCl_4^- \rightleftharpoons [H_2O^+\text{—}MCl_3^-] + Cl^- \qquad (2.14)$$

$$Nu^- + MCl_4^- \rightleftharpoons Nu\text{—}MCl_3^- + Cl^- \qquad (2.15)$$

solvent or of other environmental change will of course differ with the charge-type, as was illustrated many years ago by Brønsted (1922, 1925).

Consideration of the properties and behaviour of even such simple compounds as the hydrogen halides in different environments helps to illustrate the complexity and diversity of factors influencing thermodynamic stability and reactivity. It is apparent that the hydrogen halides can exist as diatomic molecules, strongly bonded and hence relatively stable in the gas phase. In dilute solution in aprotic solvents, or as the pure liquids, they exist in associated forms which themselves can ionise, probably in more than one way: (2.16). In hydroxylic solvents they

$$3HF \rightleftharpoons H_2F_2 + HF \rightleftharpoons H_2F^+ + HF_2^- \qquad (2.16)$$

exist partly as covalent molecules, partly as ion-pairs, and partly as dissociated ions in which the proton is powerfully solvated and probably bonded covalently to the solvent: (2.17). In acetic acid, for example, which has a low dielectric constant (*ca* 6), the dissociation to form ionic species as measured by conductivity is rather small and lies in the order $HCl < HBr < HI$; in water, these acids are completely dissociated, whereas the dissociation of hydrogen fluoride is incomplete. All the

$$HX + SOH \rightleftharpoons [SOH_2^+ \ X^-] \rightleftharpoons SOH_2^+ + X^- \qquad (2.17)$$

proton transfers to oxygen are very fast; those to halogen are probably fast also.

It is thus seen that the ionisation of a hydrogen halide in water, (2.17), is best considered as an example of the replacement process, (2.11), rather than of a heterolysis, (2.10); and that forms of ionisation more complex than that of (2.10) can arise for the hydrogen halides because the free electrophile, H^+, has a very great tendency to co-ordinate even with relatively weakly nucleophilic species. The problem of realising a true heterolysis as an elementary process, and of identifying not only the position of equilibrium but also the individual rate coefficients for its establishment, presents a separate problem for each electrophile, as we shall see when halogenations are considered.

It will be clear from the above discussion that, when covalent halides act as sources of halide anions for the formation of a new covalent bond, they often do so by first liberating the corresponding halide ions to provide a better nucleophile. That ion-pairs can in principle act as nucleophiles has been recognised for a long time (cf. Acree, 1913; Robertson and Acree, 1915; Evans and Sugden, 1949); and there is no reason in principle why a covalent halide should not provide halogen for nucleophilic combination. Clear examples exist, e.g. (2.18), in Friedel–Crafts

$$R.Cl + AlCl_3 \longrightarrow [R^+AlCl_4^-] \qquad (2.18)$$

and related reactions, where a Lewis acid assists the removal of nucleophilic halogen from an organic halide. Electrophilic catalysis by general acids (e.g. HF) of the replacement reactions of alkyl halides can also be held to provide examples.

2.7 Cationic halogen species

We have already noted that, so far, the existence of free monatomic halogen cations has not been established in solution. Cationic species in which the halogen cation is stabilised by covalent bonding with a neutral nucleophile are known, however, and these fall into several classes. The first includes the dark red compounds obtained by oxidising halogens at relatively low temperatures in highly acidic solvents (e.g., (2.19) and (2.20); Gillespie and Passmore, 1972). The method used for

$$3I_2 + S_2O_6F_2 \xrightleftharpoons{\text{in } HSO_3F} 2I_3^+ + 2SO_3F^- \tag{2.19}$$

$$3Br_2 + S_2O_6F_2 \xrightleftharpoons{\text{in } HSO_3F} 2Br_3^+ + 2SO_3F^- \tag{2.20}$$

the formation of the corresponding trichlorine cation, (2.21), illustrates that such compounds can formally be treated as halogen cations stabilised by a neutral nucleophile; the cationic charge of course becomes delocalised, and the ions are angular rather than linear. In contrast with

$$Cl_2 + ClF + AsF_5 \longrightarrow Cl_3^+ + AsF_6^- \tag{2.21}$$

the situation existing for the trihalide ions, full formal covalent bonds can be written without allowing the octet of the central halogen to be exceeded, and the covalent bond lengths are correspondingly short. Thus, as determined by X-ray crystallography, the geometry of the ICl_2^+ cation in $ICl_2^+SbCl_6^-$ is as is shown in structure (**2.19**), with an I–Cl covalent bond length of 231 pm, almost exactly as would be expected from the sum of the normal covalent radii. It is interesting also that two other

$$\left[\begin{array}{c} I \\ \diagdown \\ Cl \quad 92° \quad Cl \end{array} \right]^+$$

(**2.19**)

chlorine atoms from $SbCl_6^-$ ions are distant no more than 300 pm from the central iodine atom. This is much closer than the van der Waals touching distance, about 400 pm.

The tendency for cationic halogen to increase its co-ordination is shown also in other species involving stabilisation by neutral nucleophiles. A large number of compounds analogous to iodine dipyridine perchlorate, ((2.22) where py = pyridine, C_6H_5N), have been prepared. In the corresponding iodides, the cation has been shown (Hassel and Hope, 1960; Hope and Lii, 1970) to be centrosymmetric and planar, with linear co-ordination: structure (2.20). The N–I bond distance (216 pm)

$$I_2 + 2py + AgClO_4 \longrightarrow AgI(s) + [I, 2py]^+ClO_4 \qquad (2.22)$$

is somewhat greater than the sum of the normal covalent radii (198 pm), a result which may reflect the fact that di-co-ordination requires no-bond resonance: structures (2.21), (2.22).

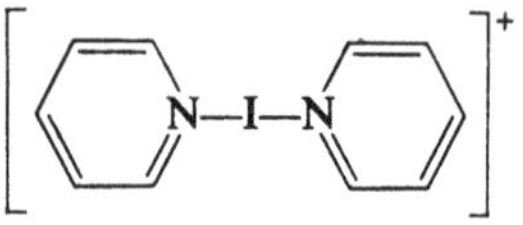

(2.20)

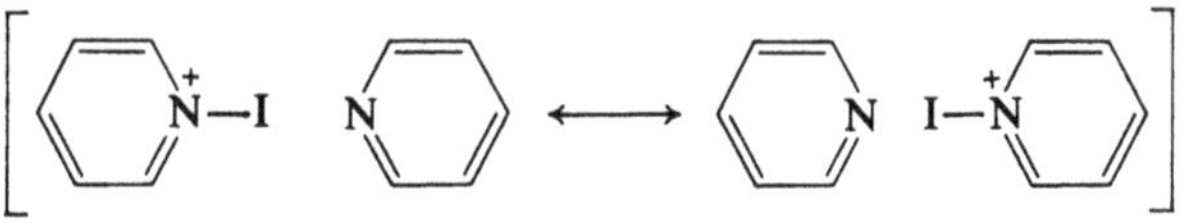

(2.21)   (2.22)

   Direct physical evidence identifying the related oxygen-containing species has probably not been provided other than through kinetic studies, though electrometric determination of the equilibrium constant for the formation of the hypo-iodous acidium ion, H_2OI^+, (2.23), has been claimed (Bell and Gelles, 1951; cf. Arotzky and Symons, 1962). Lambert *et al.* (1972) have used the results of studies of conductivity and of ^{1}H-n.m.r. spectroscopy to characterise the structures of complexes between sulphur- and selenium-containing heterocycles such as thiane and selenane and the higher halogens. Complexes of the various forms shown in structures (2.23), (2.24) and (2.25) were recognised, and strong evidence was adduced for the existence of the dissociation shown in (2.24) from changes in conductivity observed when an excess of bromine was added to a solution of thiane in ethylene dichloride.

$$I_2 + H_2O \rightleftharpoons H_2OI^+ + I^- \qquad (2.23)$$

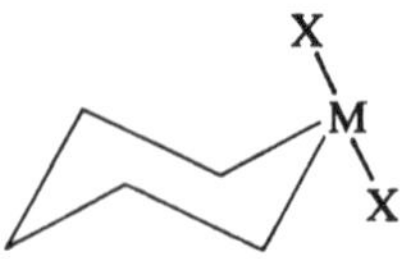

(2.23)    (2.24)    (2.25)

Other neutral species which have been thought to co-ordinate with positive halogens and thus to stabilise them in solution include the silver halide molecules: (2.25).

$$ClOH + AgCl(s) + H^+ \longrightarrow [ClAgCl]^+ + H_2O \qquad (2.25)$$

A group of cations rather different in character from the fully electron-paired compounds so far considered comprises the paramagnetic compounds exemplified by the radical-cation, $I_2^+\cdot$. Iodine in 65 per cent oleum provides a blue paramagnetic solution, which was at one time thought to contain I^+. This cation, if not stabilised by co-ordination to complete its octet, could be paramagnetic; in the six electrons to be distributed between the s- and p-orbitals of the octet, only four need to be paired, and the remainder should according to Hund's rule be allocated singly to the two highest remaining unfilled orbitals. It is now known (Gillespie and Passmore, 1972) that the species under observation is instead a radical-cation, $I_2^+\cdot$. This can be formed in a number of ways; quantitatively, for example, in fluorosulphuric acid according to (2.26).

$$2I_2 + S_2O_6F_2 \xrightarrow{\text{in } HSO_3F} 2I_2^+\cdot + 2SO_3F^- \qquad (2.26)$$

A similar reaction has been used in the super-acid solvent $HSO_3F.SbF_5 . 3SO_3$ to make $Br_2^+\cdot$.

Blue solutions of iodine in fluorosulphuric acid turn red on cooling, because of dimerisation to give the diamagnetic I_4^{2+} cation. The reaction of this with I_2 gives I_3^+, (2.27), thus providing a mechanism for the existence of the reactions (2.19) and (2.26) side by side in the same solvent.

$$I_4^{2+} + I_2 \rightleftharpoons 2I_3^+ \qquad (2.27)$$

The removal of an electron in forming $X_2^+\cdot$ from X_2 would lessen the electrostatic repulsion between the atomic nuclei, and hence allow the formation of a shorter and stronger interhalogen bond. This expectation is borne out by experiment; the Br–Br bond length in Br_2 is 228 pm, and in $Br_2^+\cdot$ is 215 pm.

The chemistry of these radical-ions has been little studied, and distinction cannot yet be made between their reactions and those of the diamagnetic species with which they are in equilibrium.

On general grounds, astatine would be expected to form the same types of compounds as are formed by iodine; and the free At^+ cation would be expected to be the most easily identified of all the possible cationic halogens. The existence of the compound $[At, py_2]^+ ClO_4^-$ has been deduced from experiments using iodine as a carrier for trace quantities of radioactive astatine; and it has been said that the best routes for synthesis of organic compounds containing C–At bonds involves keeping the halogen under oxidising conditions in which it will be kept in the cationic state. The known syntheses of astatobenzene, however ((2.28)–(2.32); Downs and Adams, 1972) would equally well be interpreted in terms of free-radical or nucleophilic processes, and further investigation of cationic astatine seems desirable.

$$Ph_2I.I \xrightarrow{\quad At^- \quad} Ph_2IAt \xrightarrow{\quad heat \quad} PhI + PhAt \qquad (2.28)$$

$$PhI \xrightarrow{\quad AtI \quad} PhAt \qquad (2.29)$$

$$AtI_2^- \xrightarrow{\quad PhNH.NH_2 \quad} PhAt \qquad (2.30)$$

$$PhN_2Cl \xrightarrow{\quad At^- \quad} PhAt \qquad (2.31)$$

$$PhI \xrightarrow{\quad At^-, \ 130\text{-}200\ ^\circ C \quad} PhAt \qquad (2.32)$$

2.8 Covalent halides as sources of electrophilic halogen

2.8.1 Introduction. We have already seen that nucleophilic halogen can be provided for the formation of new covalent bonds not only as the free halide ions or as their complexed anionic forms but also as neutral molecules. The corresponding situation applies to electrophilic halogen, appropriate change of sign being made: electrophilic halogen could be provided by the free halogen cations, if they existed; by their complexed cationic forms, the properties of some of which have been noted in the previous section; or by neutral molecules.

Species which potentially can act in this way exist in great variety, and can be classified according to the atom to which the halogen is bonded. We have already mentioned the halogen molecules and inter-halogen compounds, including those with polyligant halogen (§§2.2, 2.3). Of other compounds containing triligant halogen, the one best known as an electrophilic source is benzene iododichloride, which is formed when iodobenzene and chlorine are mixed in solution, and crystallises as a yellow solid containing dimeric units having the shape and dimensions shown in structure (**2.26**). Here the covalent I–Cl bonding distance (245 pm) is again greater than the sum of the covalent radii (227 pm). As would be expected from its structure, this compound is a chlorinating rather than an iodinating agent.

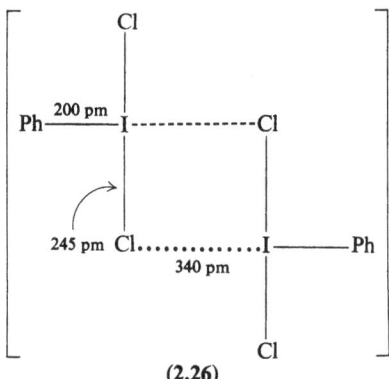

(2.26)

2.8.2   Halogen–oxygen and halogen–sulphur bonds.

Well-characterised electrophilic chlorinating agents containing Cl–O bonds include hypochlorous acid (ClOH, which is known only in solution). Attempts to distil it give only chlorine monoxide (Cl–O–Cl), formed through the equilibrium of (2.33).

$$2ClOH \rightleftharpoons Cl—O—Cl + H_2O \qquad (2.33)$$

Of the organic hypochlorites, the most stable is t-butyl hypochlorite, formed according to (2.34). Chlorine acetate ($Cl.O.CO.CH_3$) is an

$$Cl_2 + t\text{-}Bu—OH \longrightarrow t\text{-}Bu—O—Cl + HCl \qquad (2.34)$$

example of the many other derivatives of oxy-acids which are available in solution; it can be obtained, for example, by distilling the product of reaction of chlorine and mercuric acetate in acetic acid, (2.35), and can

$$Cl_2 + Hg(O.CO.CH_3)_2 \longrightarrow Hg(O.CO.CH_3)Cl + Cl.O.CO.CH_3 \qquad (2.35)$$

be characterised in solution through its spectral properties. Sulphuryl chloride ($Cl.SO_2Cl$) is also known to act as an electrophilic chlorinating agent.

Most of the corresponding bromo- and some related iodo-compounds are known, though not all of them have been characterised as fully.

2.8.3   Halogen–nitrogen bonds.

Important in this class are the N-chloro-, N-bromo- and N-iodo-amines, anilides, and sulphonamides. These, of which phenyldichloramine ($Ph.NCl_2$), N-chloroacetamide ($CH_3.CO.NHCl$), N-chloro-acetanilide ($Ph.N.(CO.CH_3)Cl$) and N-chloro-benzenesulphonamide ($Ph.SO_2.NHCl$, chloramine-T) are well-known examples, are often prepared by reaction, (2.36), of the corresponding amine or a suitable derivative with a hypohalous acid in alkaline solution.

$$Ph.NH.CO.CH_3 + ClOH \longrightarrow Ph.N(CO.CH_3)Cl + H_2O \qquad (2.36)$$

2.8.4 Halogen–carbon bonds. Among the groups of organic compounds regarded as containing 'positive halogen' are those, such as iodotrifluoromethane, for which the heterolysis of a halogen cation would leave a relatively stable carbanion (2.37) (Banus, Eméleus and Haszeldine,

$$I.CF_3 \rightleftharpoons I^+ + CF_3^- \tag{2.37}$$

1951). Another group, the members of which can be used as halogenating agents, are those which, by attachment of a proton elsewhere in the molecule, can liberate a halogen cation while leaving a resonance-stabilised organic residue; the halogenocyclohexa-2,5-dienones and their analogues, (2.38), provide examples.

$$\tag{2.38}$$

2.8.5 General considerations. In later chapters, the specific uses of most of these diverse halogen-releasing compounds will be encountered. General points arising from consideration of their possible modes of operation include the following matters. First, the exact nature of the electrophile needs to be established. Consider, for example, chlorination carried out by dissolving t-butyl hypochlorite in slightly wet acetic acid as solvent, and then adding the unsaturated compound to be chlorinated. This is quite a satisfactory procedure for some heterolytic chlorinations; among the possible chlorinating species are $Cl.OCMe_3$, $Cl.O.CO.CH_3$, $Cl.OH$, $Cl.OCl$, $Cl.OHCMe_3^+$, $Cl.OH.CO.CH_3^+$, $Cl.OH_2^+$, $Cl.OHCl^+$ and Cl^+. Detailed physical and mechanistic study is needed to establish which species are present in bulk concentration at the beginning of the reaction; which compounds are derivable from them fast enough and in sufficiently high concentration that they need to be considered as possible contributors to the chlorination; and what is the rate-limiting stage of the chlorination and what is the stoicheiometry of the transition state. If these questions can be answered, the further point arises, namely whether by change in the substrate or in the reaction conditions the mechanism can be changed so that a different reagent becomes effective.

It would help in answering some of the questions implicit in the above description if theoretical predictions could be made concerning the relative electrophilicity of halogenating species; because then a knowledge of the bulk concentrations of the individual contributors to the pre-equilibrium between reagent and solvent would allow certain of the possibilities to be ruled out. So far, however, theoretical approaches to relative electrophilicities have not been very successful, probably because the reactions which provide electrophilic halogen are usually performed in solution, are bimolecular in character, and vary in details of the mechanism. That they are performed in solution makes specific features of solvation very important; that they are bimolecular requires that bond-forming and bond-breaking forces change in a complex way with change in reagent and in substrate; and that they vary in detailed mechanism contributes to the difficulty of predicting how these changes will change the ease of reaction. The mechanism of N-chlorination, (2.36), exemplifies this in an extreme form. It has been shown kinetically that this reaction in water can involve the bimolecular reaction of the hypochlorite ion with the acetanilide molecule (Mauger and Soper, 1946); C-chlorination does not proceed under these conditions. Presumably the transition state has the form given in structure (**2.27**), in which a negative ion having a relatively strong Cl–O bond plays the unusual role of an electrophile, helped by the specific solvation of the hydroxide ion and the relative rapidity of bond breaking of the N–H bond which becomes electrophilically replaced. By contrast, for the same substrates reacting with chlorine in aqueous acetic acid, C-chlorination and N-chlorination proceed at similar rates (Orton, Soper and Williams, 1928).

$$
\begin{array}{ccc}
\underset{\displaystyle \text{Cl}\!-\!\!-\!\text{O}^-}{\overset{\displaystyle \overset{\textstyle \text{CO.CH}_3}{|}}{\text{Ph}\!-\!\underset{\curvearrowleft}{\text{N}}\!\!\curvearrowright\!\text{H}}} & \longrightarrow & \underset{\displaystyle \text{Cl}}{\overset{\displaystyle \overset{\textstyle \text{CO.CH}_3}{|}}{\text{Ph}\!-\!\text{N}}} \quad + \ \text{OH}^- \qquad (2.39)
\end{array}
$$

(**2.27**)

The complexity of problems of relative electrophilicity can be exemplified in another way. It has already been seen that for the monoligant inter-halogen compounds the relative electronegativity allows prediction of which of the two halogens acts as the electrophile in attacking an unsaturated compound. This approach cannot, however, be extended to the covalent compounds under comparison in this section. Thus R_3C–X and R_2N–X bonds will generally be polarised so that the halogen, X, bears more negative charge than does the atom to which it is attached; such compounds, however, can act as sources of electrophilic halogen rather than as sources of electrophilic carbon or nitrogen respectively. The polarisability of the Cl–X bond under the influence of a substrate capable of combining either with Cl or with X is clearly more important in determining which group acts as the electrophile than is the original direction of polarisation of the bond.

The importance of polarisability becomes further apparent when the modes of reaction of the organic hypochlorites (Cl.O.R) are compared with those of their thio-analogues, Cl.S.R. The former provide electrophilic halogen for aromatic substitution, (2.40); the latter provide electrophilic sulphur, (2.41).

$$Ar.H + Cl.OR \longrightarrow Ar.Cl + H.OR \qquad (2.40)$$

$$Ar.H + RS.Cl \longrightarrow Ar.SR + H.Cl \qquad (2.41)$$

2.9 Steric and electronic effects of halogen substituents

2.9.1 Steric effects. Fluorine is a small group, similar in effective size to hydrogen. The other halogens are considerably larger: bromine, for example, has a van der Waals radius rather similar to that of a methyl group (table 2.1) and through its bulk can be expected to have a rather similar steric effect. Thus the two bromine atoms in 2,2'-dibromobiphenyl-4,4'-dicarboxylic acid (structure (**2.28**)) introduce an activation energy for passage through the planar conformation of 79 kJ mol^{-1} as measured by the rate of racemisation of its chiral form (Harris and Mitchell, 1960). In general, as expected, chlorine is found to be effectively

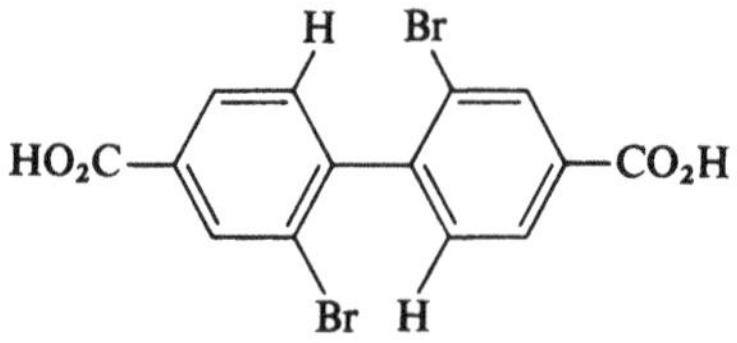

$$HO_2C \qquad \qquad CO_2H$$

(**2.28**)

less bulky than bromine, which itself is less bulky than iodine. Thus the chlorine analogue of (**2.28**) has not been resolved, but its iodine-substituted counterpart is substantially less easily racemised.

When polar and steric effects of halogens are compared with those of other groups, it should be remembered that the halogens are spherically symmetrical, so their properties are less dependent on the immediate molecular environment than are those of substituents such as OMe, CO_2H and NO_2.

2.9.2 Inductive and conjugative effects. A recent review of the directing, activating and deactivating effects of the halogens has been given by Modena and Scorrano (1973). Inductively, halogens are electron-withdrawing in character and their electron-withdrawing nature decreases with increasing size, essentially because the non-bonding electrons in the inner electronic shells increasingly shield the bonding electrons from the influence of the nuclear charge. The experimental evidence usually adduced in support of this view is derived from the strengths of the appropriately substituted organic acids, which fall in the expected order.

It has long been known, however, that this electronic influence can be modified in more than one way. In situations in which conjugative electron release can allow delocalisation of the lone-pairs of electrons from the halogen to an attached unsaturated system, the halogen appears to have an electron-releasing influence, greatest for fluorine, which can partly or wholly offset the inductive influence of the substituent. The importance of this effect is illustrated by the *ortho–para*-directing power of halogens for electrophilic aromatic substitution. This indicates the existence of some electronic influence specific to the positions conjugated with the halogen and opposed to the inductive effect, as is shown by curved arrows in structure (**2.29**), or by the canonical resonance forms in (**2.30**).

(**2.29**)   (**2.30**)

The substituent constants given for reference in table 2.5 can be used to illustrate the superiority of fluorine over the other halogens in electron release when situated conjugatively with respect to the reaction centre. Thus the combination of inductive electron withdrawal and conjugative electron release leaves chlorine, bromine and iodine almost always electron-withdrawing in an overall sense; but fluorine can in some situations be electron-releasing, as is indicated for example by the negative sign of the substituent constant, σ_p^+.

It might be expected also that a highly polarising reagent might induce electronic movements of appropriate sign, greater for the more polarisable halogens. Evidence for an inductomeric effect of this sort is perhaps to be seen through the fact that iodine, like fluorine, can sometimes be superior to bromine and chlorine in its overall power of electron release (see table 2.5).

The co-existence of these conflicting electronic influences makes a description of the relative effects of halogen a complex matter, and one

TABLE 2.5   *Substituent constants for halogens and for some other representative substituents*[a]

Substituent (R)	Electronic type[b]	σ_m[c]	σ_p[c]	σ_m^+[d]	σ_p^+[d]
MeO	−*I*, +*K*	0.12	−0.27	0.05	−0.78
HO	−*I*, +*K*	0.12	−0.37	—	−0.92
Me	+*I*, +*K*	−0.07	−0.17	−0.07	−0.31
F	−*I*, +*K*	0.34	0.06	0.35	−0.07
Cl	−*I*, +*K*	0.37	0.23	0.40	0.11
Br	−*I*, +*K*	0.39	0.23	0.41	0.15
I	−*I*, +*K*	0.35	0.18	0.36	0.14
O.CO.CH$_3$	−*I*, +*K*	0.39	0.31	—	—
O$_2$N	−*I*, −*K*	0.71	0.78	0.67	0.79

[a] Values of σ are those given by McDaniel and Brown (1958). Values of σ^+ are those given by Stock and Brown (1963).

[b] Classification is according to Ingold's terminology (§1.10). The Me group is accorded a +*K* effect because of hyperconjugation.

[c] Substituent constants derived from the ionisation constants of substituted benzoic acids (§1.11).

[d] Substituent constants derived from rates of solvolysis of aryldimethylcarbinyl chlorides, R.C$_6$H$_4$.CMe$_2$.Cl (§1.11).

which is certainly not easily disposed of by use of one or even two terms in a linear free-energy equation. A further complication comes about through the ability of the halogens to use their lone-pairs of electrons for intramolecular co-ordination with a reaction centre. Such behaviour is sometimes termed '*neighbouring-group participation*'; the term '*anchimeric effect*' is also used by some authors. The order of effectiveness of the halogens in this capacity is not that of hydrogen-bonding to external electrophilic solvent. For the latter function, the charge density around the halogen seems to be of prime importance, so that fluorine is the most and iodine the least effective of the halogens. For neighbouring-group interaction, however, fluorine is relatively ineffective, whilst chlorine, bromine and iodine are increasingly superior. Equation (2.42) represents an extreme example in which the properties of the resulting bromonium ion, (**2.31**), can be identified spectroscopically. Less powerful interaction, in which facilitation or stereochemical control of a reaction is achieved without the development of a complete covalent bond, may also be significant chemically. It is to be noted also that neighbouring-group interaction is not confined to the formation of three-membered rings; (2.43) and (**2.32**) provide an example.

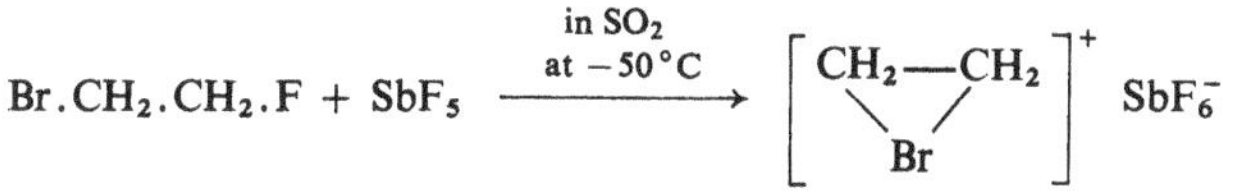

$$\text{Br.CH}_2.\text{CH}_2.\text{F} + \text{SbF}_5 \xrightarrow[\text{at} -50\,°\text{C}]{\text{in SO}_2} \left[\begin{array}{c} \text{CH}_2\!\!-\!\!\text{CH}_2 \\ \diagdown \;\; \diagup \\ \text{Br} \end{array} \right]^+ \text{SbF}_6^- \quad (2.42)$$

$$(\mathbf{2.31})$$

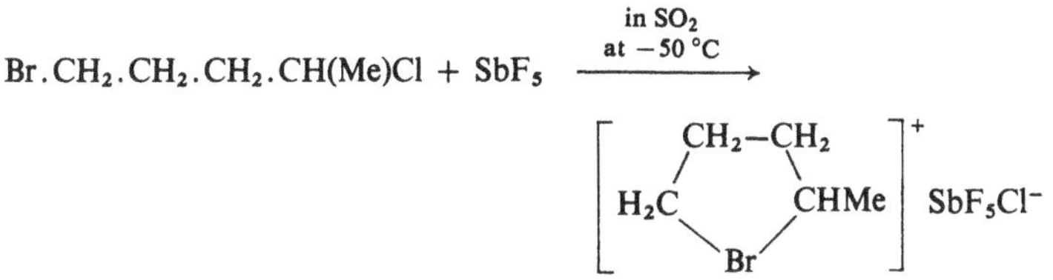

$$\text{Br.CH}_2.\text{CH}_2.\text{CH}_2.\text{CH(Me)Cl} + \text{SbF}_5 \xrightarrow[\text{at} -50\,°\text{C}]{\text{in SO}_2}$$

$$\left[\begin{array}{c} \text{CH}_2\!\!-\!\!\text{CH}_2 \\ \diagup \qquad \diagdown \\ \text{H}_2\text{C} \qquad \quad \text{CHMe} \\ \diagdown \quad \diagup \\ \text{Br} \end{array} \right]^+ \text{SbF}_5\text{Cl}^- \quad (2.43)$$

$$(\mathbf{2.32})$$

2.10   Pseudo-halogens

Several covalent compounds are known to have similarities to the halogens sufficient to merit their description as pseudo-halogens. The most important of these are cyanogen, thiocyanogen, selenocyanogen and azidocarbon disulphide, having formulae $(CN)_2$, $(SCN)_2$, $(SeCN)_2$ and $(SCSN_3)_2$ respectively. All these are volatile, and can be treated as analogous with the halogens, X_2, in that they can act as sources of a radical, X, a cation, X^+, or an anion, X^-. Consequently some or all of them can perform electrophilic additions or substitutions, just as the halogens do. Thus thiocyanogen is known to add to olefinic substances in acetic acid, being rather less reactive than iodine chloride, and so has been used to estimate dienes (which effectively add only one molecule of thiocyanogen but two of iodine chloride) in admixture with mono-enes.

3. The general patterns of reactions of electrophiles with unsaturated compounds

'...puzzling questions,...not beyond all conjecture' (Browne)

3.1 Introduction

The study of organic chemistry familiarises us at an early stage with two types of reaction of electrophiles with unsaturated compounds. The first is a reaction of addition, characteristically undergone by olefinic compounds (e.g. 1,2-addition of hydrogen chloride to propene, (3.1)).

$$CH_3.CH:CH_2 + HCl \longrightarrow CH_3.C(Cl).CH_3 \qquad (3.1)$$

The second is one of substitution, characteristic of benzenoid aromatic compounds (e.g. nitration of benzene, (3.2)).

$$C_6H_6 + NO_2^+ \longrightarrow C_6H_5.NO_2 + H^+ \qquad (3.2)$$

3.2 Reagents and substrates

This simple division becomes unsatisfactory when a wider range of reactants is considered. Among the electrophilic reagents encountered in such a survey are included those which can supply:

(*a*) electrophilic hydrogen (e.g. HCl, H.OR, H.OAc);

(*b*) electrophilic halogen (e.g. Cl_2, Cl.OR, Cl.NR$_2$);

(*c*) electrophilic oxygen or sulphur (e.g. O_3, $(SCN)_2$, RS.Cl);

(*d*) electrophilic nitrogen, phosphorus, etc. (e.g. NO_2OH, NOCl, PCl_5);

(*e*) electrophilic carbon (e.g. CH_3Br, $CH_3CO.Cl$, CBr_2);

(*f*) electrophilic metallic ions (e.g. $Tl(OAc)_3$, $Hg(OAc)_2$).

The types of unsaturated compound with which we need to be concerned are also quite diverse. Among the substrates which will be encountered, the following types are included:

(*a*) simple olefinic compounds (e.g. propene, cyclohexene);

(*b*) substituted olefinic compounds (e.g. $MeO.CH:CH_2$, $Me.C(OH):CH_2$, $CH_2:CH.CO_2H$);

(*c*) conjugated di- and poly-enes and their substituted derivatives (e.g. $CH_2:CH.CH:CH_2$, $CH_2:CH.CH:CH.CO_2H$);

(*d*) allenes (e.g. $Me.CH:C:CH_2$);

(*e*) cyclic benzenoid polyenes and their substituted derivatives (e.g. benzene, naphthalene, phenanthrene, phenol, toluene, nitrobenzene);

(*f*) conjugated heterocyclic compounds having six-membered rings (e.g. pyridine, quinoline);

(*g*) conjugated heterocyclic compounds having five-membered rings (e.g. furan, pyrrole, thiophene);

(*h*) acetylenic compounds (e.g. $Me.C\dot{:}CH$, $CH\dot{:}C.CO_2H$, $CH\dot{:}C. CH:CH_2$);

(*i*) compounds containing other types of double bond (e.g. $CH_3.CH:O$, $Ph.CH:NPh$, $Ph.N:N.Ph$, $Ph_2S:O$).

3.3   One-stage additions

There exists a substantial group of additions which are synchronous, or concerted, in character, in that both of the two new covalent bonds necessary for completion of the addition are substantially formed, but neither is completely formed, in the rate-limiting transition state. The Diels–Alder reaction, an example of which is given in (3.3), has a transition state of this kind having little polar character. There are many

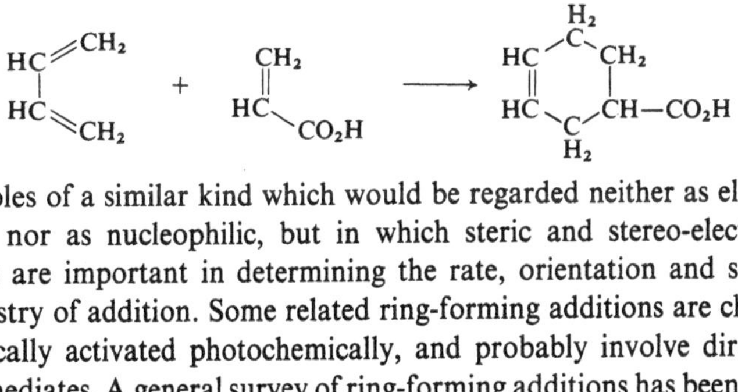

examples of a similar kind which would be regarded neither as electrophilic nor as nucleophilic, but in which steric and stereo-electronic effects are important in determining the rate, orientation and stereochemistry of addition. Some related ring-forming additions are characteristically activated photochemically, and probably involve diradical intermediates. A general survey of ring-forming additions has been given by Huisgen *et al.* (1964).

Some similar cyclic additions, however, are of a different kind, in that they are thermally activated and are markedly facilitated by the introduction of electron-releasing substituents on the double bond undergoing reaction. The combination of olefinic compounds with carbenes, (3.4),

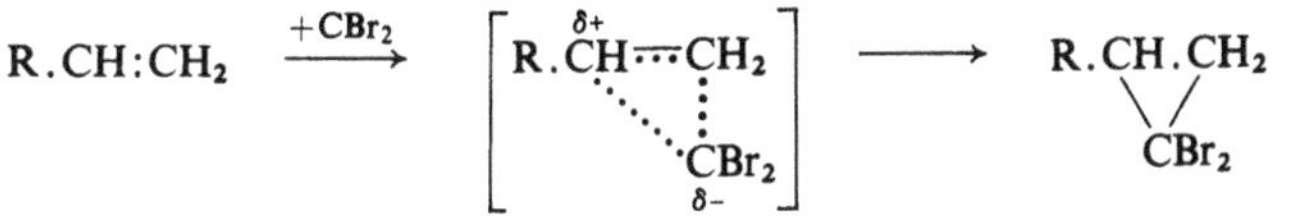

(3.1)

is a case in point. Still more marked substituent effects, described by a Hammett ρ-value of about -1 for the plot of relative rate against σ^+, are found for epoxidation: (3.5).

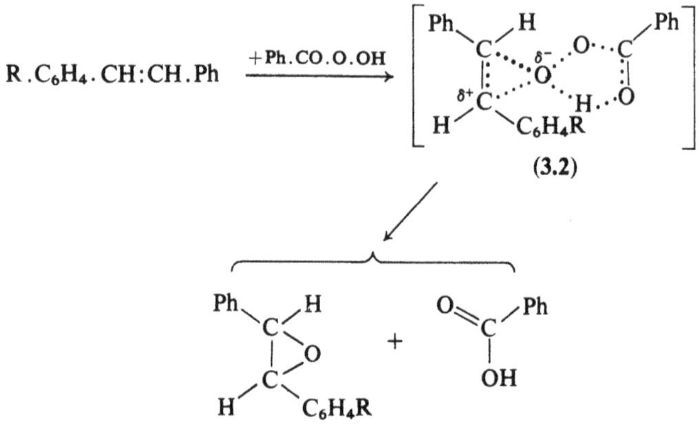

$$R.C_6H_4.CH:CH.Ph \xrightarrow{\;+Ph.CO.O.OH\;} \left[\begin{array}{c} \text{(3.2)} \end{array} \right] \tag{3.5}$$

In each of these cases, it can be deduced that the transition state (structures (3.1), (3.2) respectively) has developed dipolar character, with partial positive charge on one of the carbon atoms of the double bond, and with one of the new covalent bonds formed more strongly than the other, as is indicated in the structures by the use of strong and weak dotted lines.

These reactions can be classified as electrophilic additions through the observed influence of structure on reactivity; where reaction is completed as is indicated (e.g. (3.4) and (3.5)), it is constrained to *cis*-stereochemistry, and other side-reactions usually found for carbocationic species will either be absent, or will need conditions more forcing than those of the addition to make them occur.

Another type of concerted addition having different stereochemical consequences could occur if substrate, electrophile and nucleophile were brought together in a termolecular process: (3.6). Such reactions

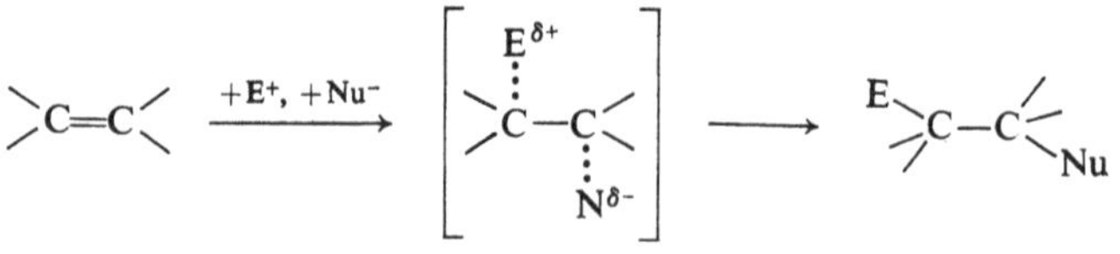

$$\diagdown C=C \diagdown \xrightarrow{\;+E^+,\, +Nu^-\;} \left[\begin{array}{c} E^{\delta+} \\ \diagdown C-C \diagdown \\ N^{\delta-} \end{array} \right] \longrightarrow \underset{\text{(3.3)}}{E} \diagdown C-C \diagdown_{Nu} \tag{3.6}$$

would be expected to prefer *trans*-stereochemistry, and the effects of substituents on the rate of reaction could indicate the dominant influence of development of a carbocationic or of a carbanionic centre, depending on the balance between the degrees to which the two reagents had become attached in the transition state. It is difficult to distinguish such cases from two-stage additions having similar transition-state stoicheiometry and hence similar kinetic form; but in considering halogenations we shall encounter some reactions the mechanisms of which must approximate quite closely to this case.

3.4   One-stage substitutions

In principle, one-stage electrophilic substitutions are also possible; and they could adopt several forms, of which three will now be outlined. In the first, one electrophile is replaced by another without the kinetic intervention of a second reagent to form a new covalent bond with the departing electrophile. The transition state is then as shown in structure (**3.4**) for the example of a generalised aromatic replacement, (3.7). Such a

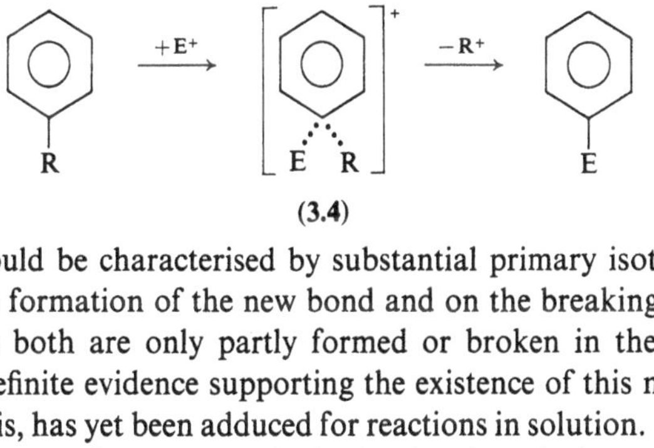

$$\qquad\qquad (3.7)$$

(**3.4**)

reaction would be characterised by substantial primary isotope effects both on the formation of the new bond and on the breaking of the old bond, since both are only partly formed or broken in the transition state. No definite evidence supporting the existence of this mechanism, simple as it is, has yet been adduced for reactions in solution.

A second type of one-stage substitution involves the intervention of a nucleophilic reagent capable of helping to remove the departing electrophile, (3.8); the transition state then takes the form shown in structure

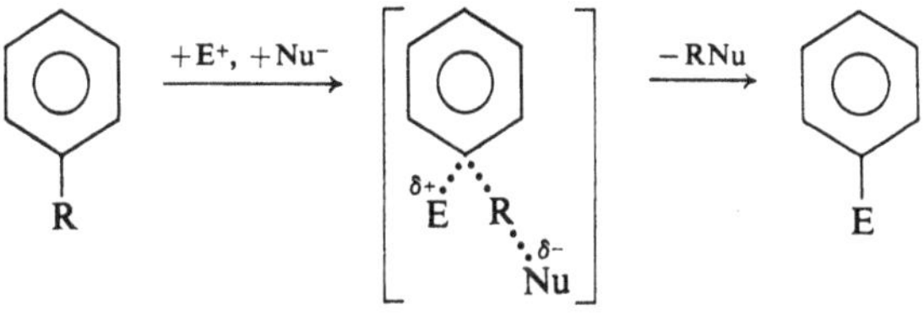

$$\qquad\qquad (3.8)$$

(**3.5**)

(**3.5**). Such a mechanism would again be characterised by substantial primary isotope effects; the kinetic intervention of the nucleophile Nu^- would also be observed, but might be difficult to establish if the solvent played the part of the nucleophile.

Some electrophilic substitutions partly meet these requirements in that the leaving-group isotope effect has been established but the forming-group isotope effect has not been investigated. Most of these are conventionally assigned to the group of two-stage reactions discussed in §3.5, and it is commonly held (cf. Taylor, 1972, p. 3) that no examples of the one-stage mechanism have yet been identified. Perhaps some intramolecular electrophilic rearrangements will in due course be shown to provide examples; one of the mechanisms available for the nitrosamine rearrangement could be of this kind. Here the electrophile is supplied intramolecularly; the existence of a primary deuterium isotope effect, $k_H/k_D = 1.7$, for the formation of p-nitroso-N-methylaniline from the cation (3.6) (Williams, 1972) shows that proton loss forms part of the rate-determining step; and it can be considered that a water molecule provides the nucleophile removing the proton, thus providing a transition state such as is indicated in structure (3.7).

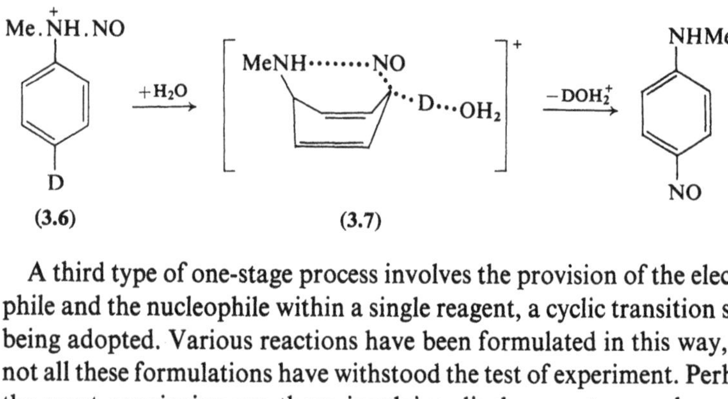

A third type of one-stage process involves the provision of the electrophile and the nucleophile within a single reagent, a cyclic transition state being adopted. Various reactions have been formulated in this way, but not all these formulations have withstood the test of experiment. Perhaps the most convincing are those involving displacement or exchange of electrophilic mercury. Equation (3.10) gives an example thought to

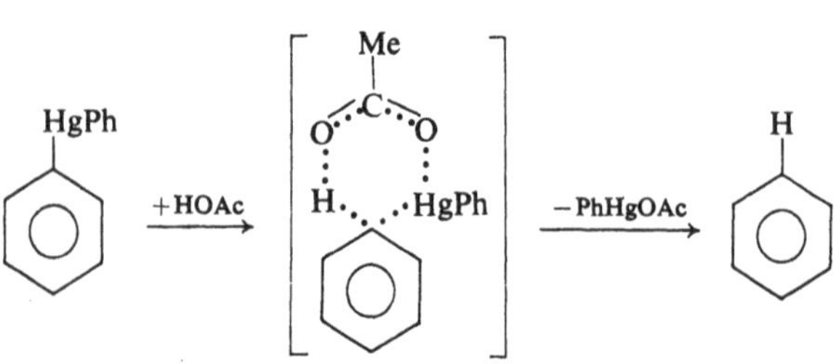

involve a cyclic six-centred transition state, structure (3.8) (Taylor, 1972, p. 280). Cyclic transition states of related kinds have been proposed for a number of other aromatic replacements, as for example for nitro-desilylations, (3.11) (Taylor, 1972, p. 378).

$$Ar.SiMe_3 + NO_2.X \longrightarrow Ar.NO_2 + R_3Si.X \qquad (3.11)$$

3.5 Two-stage reactions

3.5.1 General considerations. Much more common than any of these possible one-stage mechanisms are the two-stage processes in which an electrophile attacks the unsaturated compound to give an intermediate carbocation stable enough to have appreciable life under the conditions of reaction: (3.12). For addition reactions, the intervention of carbocations as intermediates is deduced in part from the isolation of model

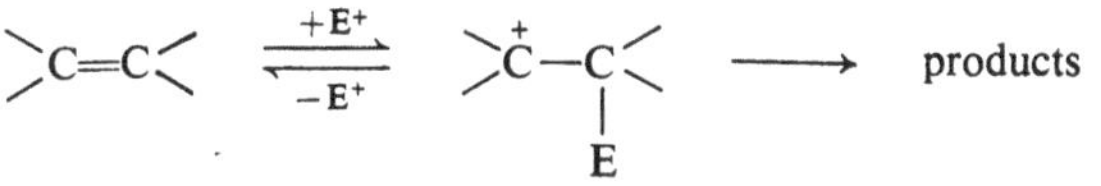

$$\begin{array}{ccc} \ce{>C=C<} & \underset{-E^+}{\overset{+E^+}{\rightleftharpoons}} & \ce{>\overset{+}{C}-C<} \\ & & \underset{E}{|} \end{array} \longrightarrow \text{products} \qquad (3.12)$$

$$(3.9)$$

carbocations as stable entities under special conditions, and in part from recognition of their typical reactions, particularly those involving diversion of the product through capture of the intermediate by an added nucleophile (for a review, see de la Mare and Bolton (1966), from which a number of the examples given in this chapter have been abstracted). Thus the reaction of 2,4-dinitrobenzenesulphenyl chloride (ArS.Cl) with *p*-methoxystyrene in acetic acid gives not only the product of addition of ArS.Cl but also that of intervention of the solvent ((3.13); here we are not specifying the detailed structure of the intermediate (**3.10**)).

$$MeO.C_6H_4.CH:CH_2 \xrightarrow[-Cl^-]{+ArS.Cl} [MeO.C_6H_4.\overset{+}{C}H.CH_2(SAr)]$$

$$(3.10)$$

$$\downarrow {\scriptstyle +HOAc, -H^+} \qquad (3.13)$$

$$MeO.C_6H_4.CH(OAc).CH_2(SAr)$$

For substitution, rather different considerations have been adduced, based particularly on studies of primary kinetic isotope effects. Melander (1950) showed for the nitration of toluene and for some other aromatic substitutions that hydrogen and its isotopes were displaced at the same rate. Since this implies that the transition state is reached without appreciable stretching of the C–H bond, it has consequently been accepted that an intermediate carbocation must be implicated. For such aromatic substitutions as show a primary isotope effect, the fact that the magnitude of this effect can often be changed by changing the nucleophile and so changing the rate of the product-forming step has been taken as strong evidence that the two-stage carbocationic mechanism is of wide generality, and this is supported by direct physical evidence for the existence of intermediates of the type required. Summaries with more details have been given by Berliner (1964), by Zollinger (1964), by de la Mare (1971), and in other textbooks dealing with reaction mechanisms.

The simple idea conveyed in (3.12) allows many further ramified and complicated chemical consequences, which depend on the natures of the electrophile and of the organic substrate as well as on the conditions of reaction. Many of these will be encountered through the later chapters of this book and here only a few of the more general questions will be outlined.

3.5.2 Kinetics and products. First, it is apparent that either the formation of the carbocation or its destruction can be rate determining; so the overall rate may be quite independent of the product-forming stages; or alternatively one or more of the product-forming stages may, when altered, change the overall rate of disappearance of the olefinic compound. Associated with these kinetic features is the possibility that the conditions or the structure of the intermediate may be such as to allow the build-up of the intermediate to high concentrations, and hence to make possible its direct detection or its isolation.

3.5.3 Orientation. Secondly, the orientation (*regiospecificity* is a term now used by some writers) of attack on the double bond becomes an important question for any substituted olefinic compound. For a compound containing only one double bond, consideration must be given to the possible consequences of isomerism among forms such as (**3.11**)–(**3.16**). These may show themselves in the extent to which the substituents attached to the double bond affect the rate of consumption of the reagents, since only if the bridged forms (**3.12**) and (**3.15**) are dominant will all four groups affect the rate according to the same laws. The nature and stereochemistry of the products will also be influenced by the existence and energetics of the isomerism: further reaction through (**3.11**) normally gives a structural isomer different from that expected through (**3.13**), and only through dominance of the bridged forms (**3.12**) will the relative geometry of R and R′ be maintained from reactant to product.

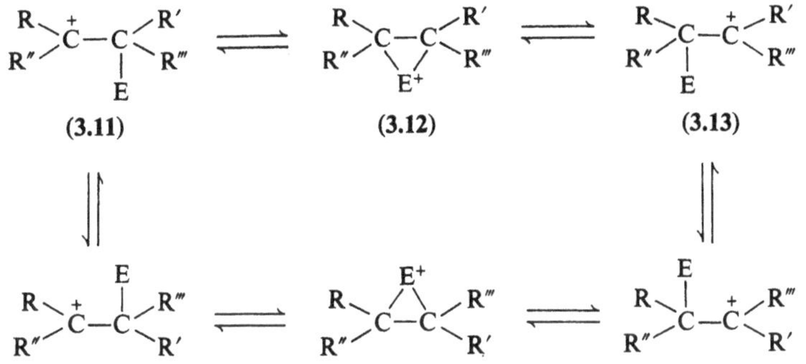

$$(3.11) \rightleftharpoons (3.12) \rightleftharpoons (3.13)$$

$$(3.14) \rightleftharpoons (3.15) \rightleftharpoons (3.16)$$

For compounds containing more than one double bond, the problems of orientation become multiplied. In the reactions of butadiene with electrophiles, for example, the orientation of initial attack is usually on the terminal carbon atom, but completion of addition can take place either at the 2- or at the 4-carbon atom, as in (3.14), to give addition

$$CH_2\!:\!CH.CH\!:\!CH_2 \xrightarrow{\ +H^+\ } [CH_3.CH\!\cdots\!CH\!\cdots\!CH_2]^+ \xrightarrow{\ +Cl^-\ }$$

(3.17) (3.14)

$$CH_3.CH(Cl)CH\!:\!CH_2 \quad \text{and} \quad CH_3.CH\!:\!CH.CH_2Cl$$

with or without double-bond rearrangement. For aromatic compounds, several positions are possible for electrophilic attack, and some or all of them may contribute to the product-forming stages. Thus the nature of the products obtained for the nitration of toluene suggests that the intermediates (3.18), (3.19) and (3.20), involving attack initially on *ortho*-, *meta*- and *para*-positions respectively, are formed, decomposing then irreversibly by proton-loss; attack on the *ipso*-position, which bears the substituent, to give the intermediate (3.21), has commonly been neglected, though its potential importance is now recognised (Hartshorn, 1974).

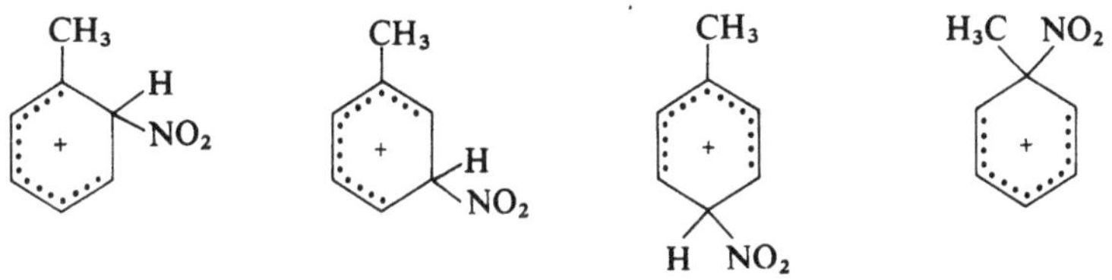

(3.18) (3.19) (3.20) (3.21)

3.5.4 Reactions of the intermediate: combination with a nucleophile.
Capture of the carbocationic intermediate by a nucleophile gives a product of addition, as in the example of (3.13). When an extended

conjugated system is under examination, double-bond rearrangement may contribute to the product mixture, as is shown for butadiene in (3.14). When the nucleophile is provided intramolecularly, a ring closure may occur, as in the chlorolactonisation shown in (3.15).

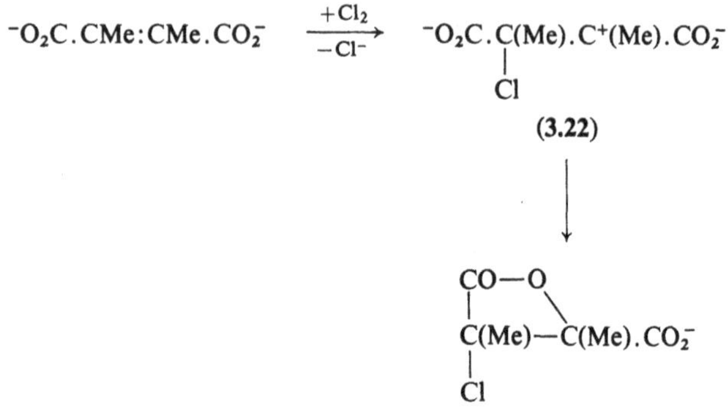

$$^-O_2C.CMe:CMe.CO_2^- \xrightarrow[-Cl^-]{+Cl_2} \; ^-O_2C.\underset{\underset{Cl}{|}}{C}(Me).C^+(Me).CO_2^-$$

$$(3.22)$$

$$(3.15)$$

$$\begin{array}{c} CO-O \\ | \qquad \backslash \\ C(Me)-C(Me).CO_2^- \\ | \\ Cl \end{array}$$

3.5.5 Reactions of the intermediate: proton loss, or loss of another electrophile.

Proton loss from the carbocationic intermediate gives a product of substitution, an example being given in (3.16). The corresponding loss of an electrophile other than a proton gives the product

$$Me_2C:CH_2 \xrightarrow[-OAc^-]{+NO_2.OAc} Me_2C^+.CH_2.NO_2 \xrightarrow{-H^+} Me_2C:CH.NO_2 \quad (3.16)$$
$$(3.23) \qquad\qquad \text{and other products}$$

of a generalised replacement, as in protodesilylations, (3.17), and related reactions (Eaborn, 1960). When a proton, or another electrophile, is lost

$$R_2C:CH.SiMe_3 \xrightarrow{+H^+} R_2C^+.CH_2.SiMe_3 \xrightarrow{-SiMe_3^+} R_2C:CH_2 \quad (3.17)$$
$$(3.24)$$

from a carbon or other atom attached to the double bond, the result is a product of replacement with rearrangement: (3.18).

$$Me_2C:CH_2 \xrightarrow[-Cl^-]{+Cl_2} Me_2C^+.CH_2Cl \xrightarrow{-H^+} CH_2:C(Me).CH_2Cl \quad (3.18)$$
$$(3.25)$$

3.5.6 Reactions of the intermediate: rearrangement by group migration.

Rearrangement by group migration within the carbocationic intermediate results in the formation of a rearranged carbonium ion, from which rearranged products of addition or group replacement can be derived. Two examples are given in (3.19) and (3.20).

$$CH_2\text{:}CH.CH_2.Br \xrightarrow{\ +H^+\ } CH_3.CH^+.CH_2Br \longrightarrow$$

(3.26)

$$CH_3.CH(Br).CH_2^+ \xrightarrow{\ +H_2O,\ -H^+\ } CH_3.CH(Br).CH_2OH \quad (3.19)$$

(3.27)

$$Me_3C.CH\text{:}CH_2 \xrightarrow{\ +H^+\ } Me_3C.CH^+.CH_3 \longrightarrow$$

$$Me_2C^+.CHMe_2 \xrightarrow{\ +Cl^-\ } Me_2C(Cl).CHMe_2 \quad (3.20)$$

(3.28)

3.5.7   Other reactions of intermediates or products.

Reactions (3.15)–(3.20) can all be regarded as involving intermediates analogous to a single one (e.g. (3.11)) of a family of intermediates, (3.11)–(3.16). In the formulations of the equations (3.15)–(3.20), the detailed structures of the intermediates (3.22)–(3.28) have not been specified; and it is not meant to imply that the reactions shown are the only ones occurring under the conditions which lead to the products shown. Intermediates or products may be unstable in other ways under the reaction conditions, and hence for individual reactions can provide further pathways leading to other types of product. Several examples of this type of behaviour are worth noting. The first is the case in which an addition is followed by a rapid elimination either under the conditions of reaction or under the conditions under which the product is worked up for analysis. Sequences of this kind may lead to the identification of apparently normal or, alternatively, of obviously abnormal products, as in the apparent electrophilic acetoxylations which accompany some nitrations in acetic anhydride, and actually involve electrophilic addition of nitronium acetate followed by elimination of nitrous acid: (3.21).

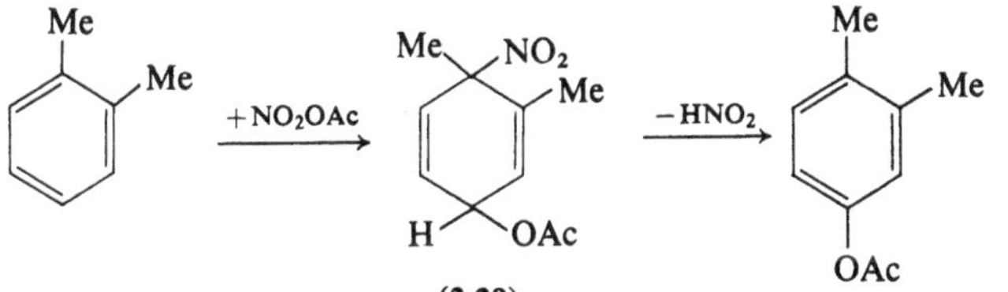

(3.29)   (3.21)

A second type of product instability occurs when the electrophile gives an initial product, the reactivity of which is similar to, or greater than, that of the starting material. The result under these circumstances is the consumption of more than one molecule of electrophilic reagent per molecule of unsaturated compound, as in the addition, (3.22), of chlorine to naphthalene.

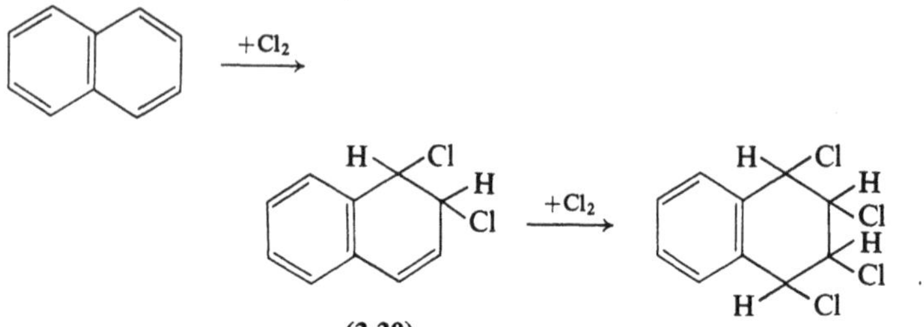

(3.30)   (3.22)

A third type of product instability can be exemplified by the sequence generally accepted for ozonolysis, where the initial product, the detailed structure of which is probably still uncertain, undergoes carbon–carbon bond fission: (3.23).

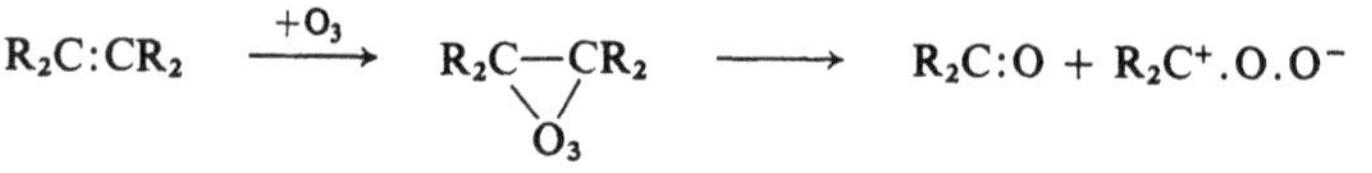

(3.31)

3.5.8 Reversibility of reaction paths. If one or more of the products of reaction is formed reversibly, the establishment of kinetic control of the reaction becomes important if control of the reaction products is required. Furthermore, the intervention of a particular reaction path giving a particular product reversibly may be vital to the determination of the overall course of the reaction; alternatively, it may be unimportant in that the product formed reversibly may merely provide a temporary repository for the reactants. The same difficulty applies to the identification of intermediates leading to particular products of reaction. Consider, for example, the nitration of toluene, giving *ortho-*, *meta-* and *para-*nitrotoluenes through intermediates (**3.18**), (**3.19**) and (**3.20**) respectively. The isomeric *ipso-*intermediate (**3.21**) also would be expected to be formed; to what extent does its formation contribute to the production of *ortho-*nitrotoluene? How could it be proved that it is not a participant in *ortho-*nitration? At the present, such questions can be answered only arbitrarily, or by the exercise of personal judgement; and few chemists would deny that from time to time their personal judgement on matters of this kind had later been proved experimentally to be wrong.

3.5.9   Intermediates isomeric with the reactants: molecular complexes, σ-complexes and ion-pairs. There are several ways in which the nucleophile as well as the electrophile can be involved in electrophilic reactions proceeding through an intermediate. One of these involves the formation of a loose complex between the two reactants; the second involves the formation of a more strongly bound σ-complex between the reagent E–Nu and the unsaturated compound without the heterolytic fission of the E–Nu bond; and the third allows that this bond has cleaved, but that the nucleophile is still associated with the cation in a discrete intermediate, an *ion-pair*. These various stages of association are illustrated in structures (**3.32**), (**3.33**) and (**3.34**).

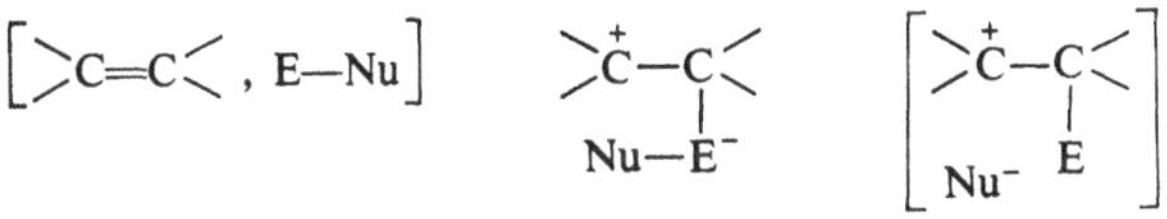

(3.32)	(3.33)	(3.34)
Molecular complex	σ-Complex	Ion-pair

In Chapter 2, structural evidence illustrating the possible importance of molecular complexes and of σ-complexes in halogenation of unsaturated compounds was summarised. The necessity of including ion-pairs also comes from studies of nucleophilic substitution, where the chemistry of reactions involving carbocationic intermediates produced by heterolysis requires the intervention of such ionic aggregates. Identification of the particular aggregates involved in specific examples is, however, not without difficulty (cf. Ingold, 1969; de la Mare and Swedlund, 1973).

3.6   Generalised pattern of reactions between electrophilic reagents and unsaturated compounds

Figure 3.1 gives a generalised pattern of reaction paths summarising the most important features outlined in this chapter as they apply to reactions between electrophilic reagents, E–Nu, and compounds containing carbon–carbon multiple bonds. Extensions to reactions involving other types of multiple bond follow the same general principles, but points of detail become modified, sometimes in ways important to the chemistry (see chapter 10). That there are so many potential complications and ramifications should not be allowed to obscure the overall generalisation

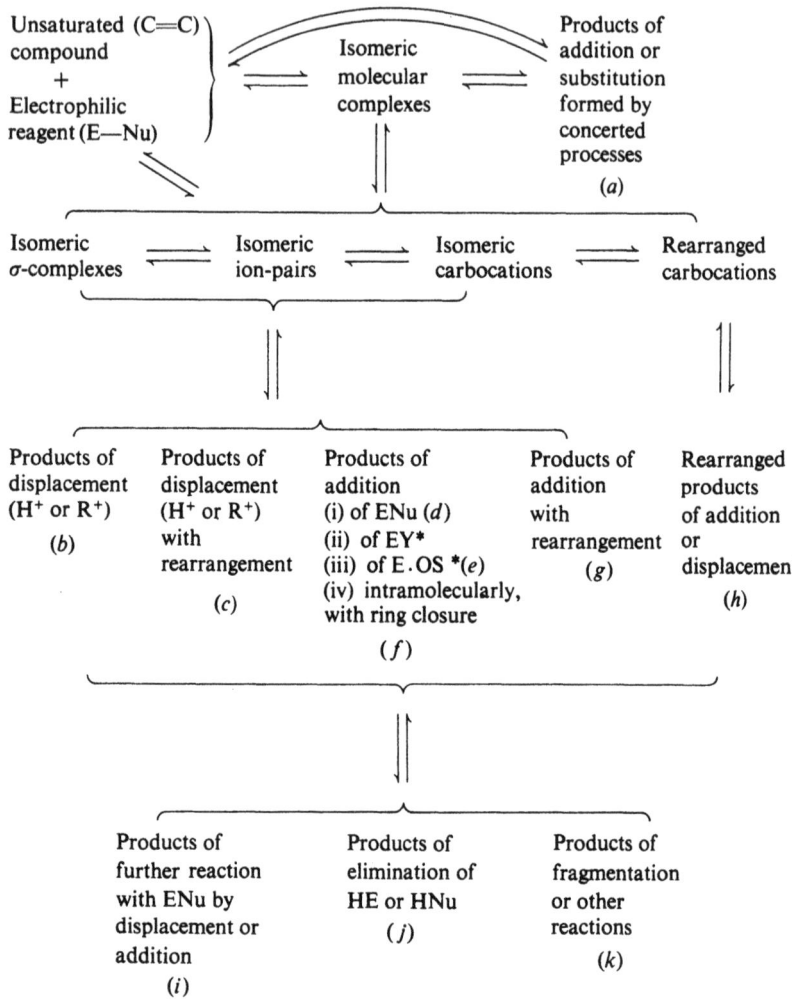

Fig. 3.1. Patterns of behaviour in reactions of unsaturated compounds with electrophilic reagents. *Y = external nucleophile; SOH = hydroxylic solvent.

   Corresponding equations in the text: (*a*) (3.3)–(3.11); (*b*) (3.2), (3.16) and (3.17); (*c*) (3.18); (*d*) (3.1); (*e*) (3.13); (*f*) (3.15); (*g*) (3.14); (*h*) (3.19) and (3.20); (*i*) (3.22); (*j*) (3.21); (*k*) (3.23).

that the reactions very frequently involve intermediates having carbocationic character, and that it is the detailed chemistry of these intermediates that determines the paths adopted, and the products formed.

4 Fluorine and its compounds as electrophiles

'...tamed and brideled the furious rage...' (*O.E.D.* vol. XI, p. 6)

4.1 Reactions of molecular fluorine

The fluorine–fluorine bond in F_2 is abnormally weak, probably because of electrostatic repulsion between the two rather poorly shielded atomic nuclei. As a result, elementary fluorine at room temperature tends to react with organic compounds by violent free-radical chain processes. Recently, however, largely through the work of Merritt and his co-workers (for a summary, see Sheppard and Sharts (1969)), it has been found possible to moderate and alter the course of the reaction by working in solvents at low temperatures. Under these conditions, a number of substrates give products characteristic of the general pathways adopted when an electrophile attacks an unsaturated compound. Thus for 1,1-diphenylethylene, addition and substitution have been shown to occur concurrently: (4.1). Free-radical reactions involving this substrate

$$\text{Ph}_2\text{C:CH}_2 \xrightarrow[\text{in FCCl}_3,\ -78\ ^\circ\text{C}]{F_2} \begin{cases} \text{Ph}_2\text{CF.CH}_2\text{F} & (14\%) \\ \text{Ph}_2\text{C:CHF} & (78\%) \\ \text{Ph}_2\text{CF.CHF}_2 & (8\%) \end{cases} \qquad (4.1)$$

usually give rearranged products, and these were absent from the products of fluorination under the conditions used. Further evidence that free-radical pathways can be avoided came from the stereochemistry of addition both to cyclic and to acyclic olefinic compounds. Thus free-radical addition of chlorine to phenanthrene (4.1) was found to give almost exclusively the *trans*-dichloride, whereas heterolytic addition gave a mixture containing a preponderance of the *cis*-isomer. When

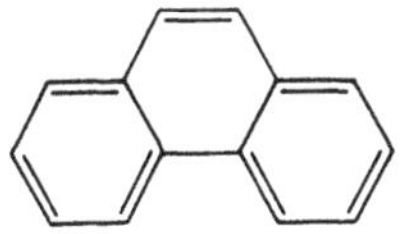

(4.1)

55

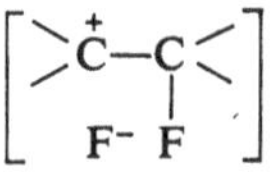

$$\left[>\overset{+}{\underset{\underset{F^-}{|}}{C}}-\overset{}{\underset{\underset{F}{|}}{C}}< \right]$$

(4.4)

Stereochemistry of the type illustrated in fig. 4.1 has been helpful generally in defining the reaction paths adopted in halogenations, and further examples will be encountered in later chapters. Other olefinic fluorinations displaying the characteristics of heterolytic processes include additions to extended conjugated systems with consequent double-bond rearrangement, as in the example shown in (4.3).

$$CH_2{=}\!\!\!\left\langle\!\!\!\bigcirc\!\!\!\right\rangle\!\!\!{=}CH_2 + F_2 \longrightarrow F.CH_2{-}\!\!\!\left\langle\!\!\!\bigcirc\!\!\!\right\rangle\!\!\!{-}CH_2F \quad (4.3)$$

In a related study of the reactions of simple aromatic compounds in the reactant as solvent, or in solution in acetonitrile, Grakauskas (1970) found that the orientation of substitution resembled that of typical electrophilic replacements. Thus toluene gave about 90 per cent of *o*- and *p*-fluorotoluenes; nitrobenzene gave about 80 per cent of *m*-fluoronitrobenzene; and naphthalene gave about 75 per cent of 1-fluoronaphthalene. For none of these fluorinations has any kinetic study yet been possible; but Grakauskas (1970) noted that the fluorination of toluene was rapid, but that of 2,4-dinitrotoluene was sluggish. For reasons both of orientation and reactivity, therefore, he suggested that the fluorination involved the usual type of carbocationic intermediate (e.g. structure **(4.5)** and its analogues).

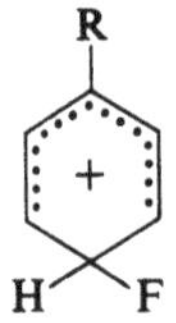

$$\underset{\overset{}{\underset{H \quad F}{\bigcirc}}}{\overset{R}{\mid}}$$

(4.5)

4.2 Electrophilic fluorination with fluoroxy-compounds

Another method of providing electrophilic fluorine while avoiding the incursion of chain reactions involving free radicals involves the use of fluoroxy-compounds. Barton *et al.* (1968, 1972) have exemplified a number of the reaction paths available when hypofluorites such as $F.OC(CF_3)_3$, $(FO)_2CF_2$ and $F.OCF_3$ are used. The reactions were found not to be affected by the presence of radical-scavengers such as oxygen or ethers, so probably they are heterolytic in character. Electrophilic replacement of a proton is shown in (4.4). Equation (4.5) shows

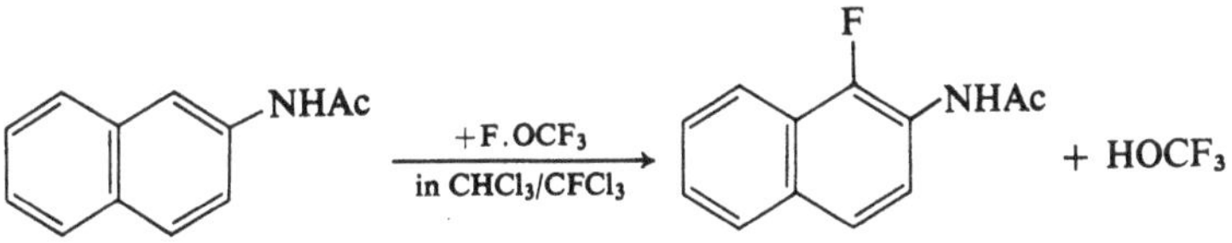

$$+ \text{ HOCF}_3 \tag{4.4}$$

electrophilic displacement of Me^+ or Ac^+ with double-bond rearrangement, and the analogous reaction in a system having extended conjugation is shown in (4.6). Electrophilic addition to the 5,6-double bond, with

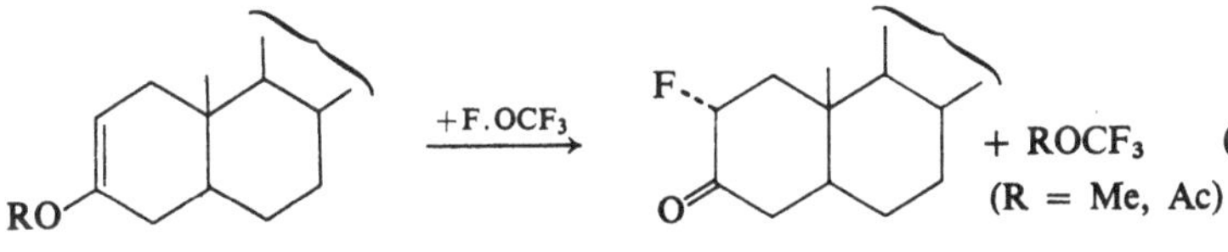

$$+ \text{ ROCF}_3 \qquad (4.5)$$
$$(R = Me, Ac)$$

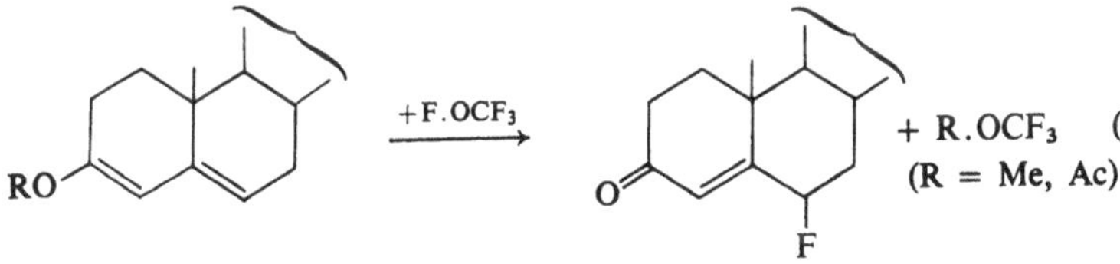

$$+ \text{ R.OCF}_3 \qquad (4.6)$$
$$(R = Me, Ac)$$

accompanying capture of fluoride ion, accompanied the last reaction; similar additions were established for other substrates (e.g. with benzofuran: (4.7)). Subsequent further electrophilic fluorinations of the

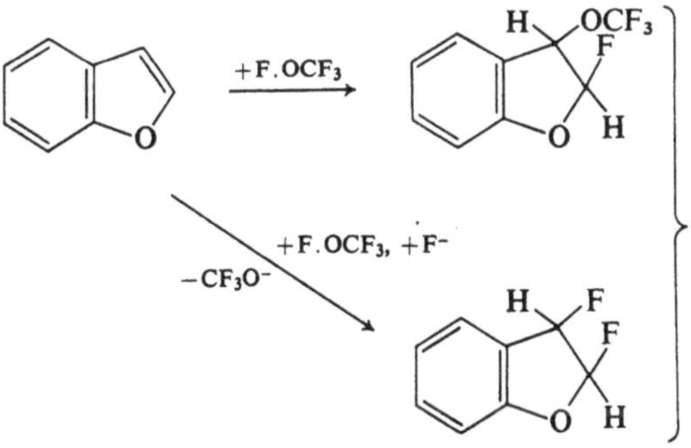

$$\tag{4.7}$$

products of addition or substitution were recognised for some of the substrates, as in the formation, (4.8), of the tetrahydronaphthalene, (**4.6**), from *N*-acetyl-α-naphthylamine and $F.OCF_3$ in $CFCl_3$ containing ethanol. Although kinetic studies of these reactions have not been

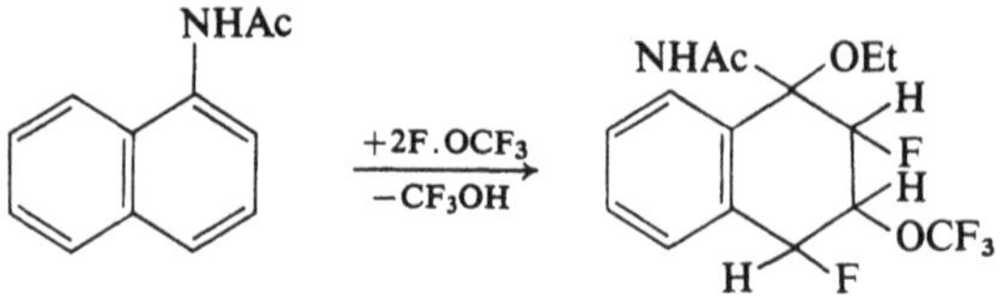

$$\text{(4.6)} \qquad \qquad \text{(4.8)}$$

performed, and so the details of the mechanistic pathways cannot be regarded as settled, it is entirely acceptable from the nature of the products that these fluorinations are electrophilic in character.

4.3   Xenon difluoride as a fluorinating agent

Xenon difluoride has the structure F–Xe–F; in solution, there is some evidence that it can dissociate under the influence of hydrogen fluoride according to (4.9): (Bartlett and Sladky, 1968). Shaw *et al.* (1969, 1970,

$$XeF_2 + HF \rightleftharpoons XeF^+ + HF_2^- \qquad (4.9)$$

1971) have shown that xenon difluoride reacts with benzene and its derivatives in carbon tetrachloride as solvent to give major amounts of fluoro-derivatives. The orientation observed in the reactions is that expected for an electrophilic substitution; thus fluorobenzene gives mainly *p*-difluorobenzene; biphenyl gives 2- and 4-fluorobiphenyl, with very little of the 3-substituted isomer; and nitrobenzene gives mainly *m*-fluoronitrobenzene. Wholly free-radical pathways, therefore, are quite improbable for reaction under these conditions; such reactions with nitrobenzene give very little *meta*-substitution (Williams, 1960). Accompanying the products of substitution of benzene, however, derivatives of biphenyl are formed in minor proportions, and radical-ions derived from biphenyl have been identified in the reacting mixtures by the use of electron spin resonance spectroscopy. The mechanism favoured by Shaw *et al.* (1971) involves a precursor common to both types of product (fig. 4.2). It is suggested that the first complex, (4.7), involves the aromatic compound, xenon difluoride, and the catalyst, hydrogen fluoride. The function of this catalyst, which seems to be essential for the fluorination, is probably to polarise the reagent; hydrogen chloride under similar conditions is ineffective, though it can become incorporated into the products at a later stage in the process. The further reaction of the intermediate, which is shown as involving loose bonding to xenon but instead could be formulated with a carbon–xenon or carbon–fluorine bond, gives by alternative concurrent modes of decomposition either the

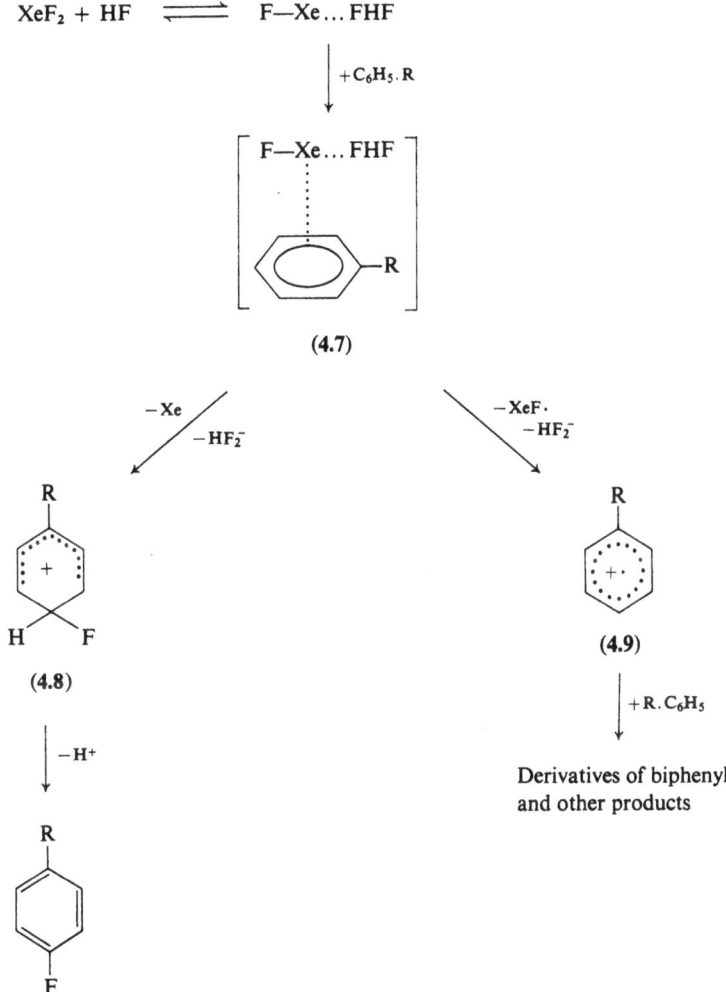

Fig. 4.2. Possible reaction pathway in the *para*-fluorination of a substituted benzene, $C_6H_5.R$, by xenon difluoride.

carbocationic intermediate (**4.8**) or the radical-ion (**4.9**). These then react further, each giving its own characteristic products.

The same reagent has been used under more polar conditions involving trifluoroacetic acid as solvent to give, (**4.10**), products of addition to

$$Ph_2C:CHR + XeF_2 \xrightarrow[\text{in } CF_3CO_2H]{-Xe} Ph_2CF.CFHR \qquad (4.10)$$

olefinic compounds (Zupan and Pollak, 1973). Small amounts of the products of interception, (**4.10**), or rearrangement, (**4.11**), of a carbo-cationic intermediate were also identified in the products.

$$Ph_2C(O.COCF_3).CFHR \qquad\qquad Ph.CO.C(Ph)HR$$

$$\textbf{(4.10)} \qquad\qquad\qquad\qquad \textbf{(4.11)}$$

The reaction was very slow in the absence of hydrogen fluoride or of trifluoroacetic acid; it clearly is similar to those involving aromatic systems, but whether it can involve radical-cations has not been established.

Related reaction pathways have been described for fluorinations involving the higher fluorides of xenon (e.g. XeF_4), which expectedly are still more active fluorinating agents, the reactions of which probably have greater tendency to provide free radicals.

4.4   Fluorination by metallic ions

Many higher metallic fluorides (e.g. PbF_4) are known to act as fluorinating agents for unsaturated compounds. The mechanisms of these reactions are not well understood; some of them may involve electrophilic metallation followed by nucleophilic replacement of the metallo-substituent. It has been suggested, however (Burdon, Parsons and Tatlow, 1972) that fluorination by the higher fluorides of some transition metals (e.g. by AgF_2, CoF_3, MnF_3 or CoF_4) can proceed by sequences involving radical-cations from which fluorine-substituted cationic intermediates of the usual type, (**4.8**) in fig. 4.2, are formed and thence lead to products of electrophilic fluorination. The sequence of reactions proposed is that of (4.11)–(4.13). The reactions have the orientational characteristics of

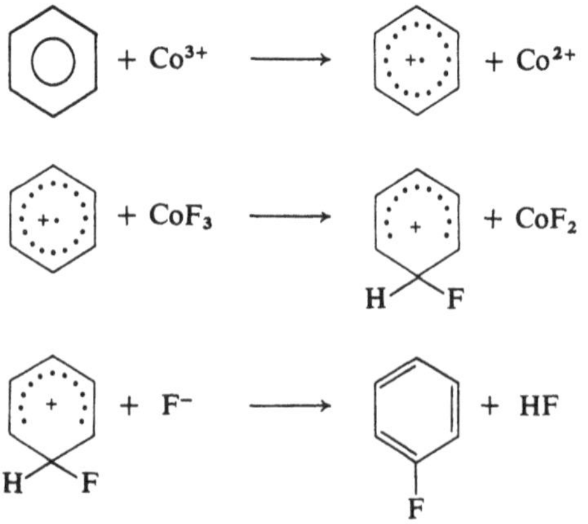

$$+ \; Co^{3+} \longrightarrow \qquad + \; Co^{2+} \qquad (4.11)$$

$$+ \; CoF_3 \longrightarrow \qquad + \; CoF_2 \qquad (4.12)$$

$$+ \; F^- \longrightarrow \qquad + \; HF \qquad (4.13)$$

electrophilic substitutions; thus fluorobenzene gives *p*-difluorobenzene; phenol gives 2,4,6-trifluorophenol; and thiophene gives 2-fluorothiophene.

4.5 Perchloryl fluoride

Perchloryl fluoride, $FClO_3$, is a tetrahedral molecule containing an F–Cl bond. Its reactions have been reviewed by Khutoretskii *et al.* (1967); and it has been described as an ambident electrophile, since with Friedel–Crafts catalysts it effects electrophilic entry of the ClO_3 group into the aromatic molecule, whereas fluorinations which are apparently electrophilic in character can supervene with sufficiently reactive unsaturated substrates. Thus substitutions in phenols have been reported by several investigators, (4.14); and the formation of 2-fluoroindanone from indene has been considered to follow the sequence shown in (4.15).

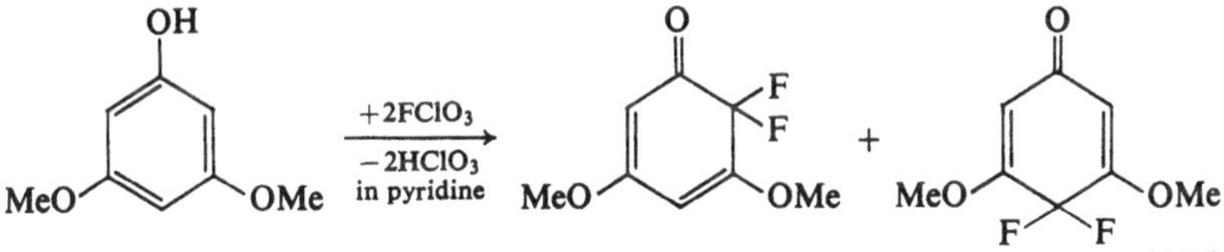

(4.14)

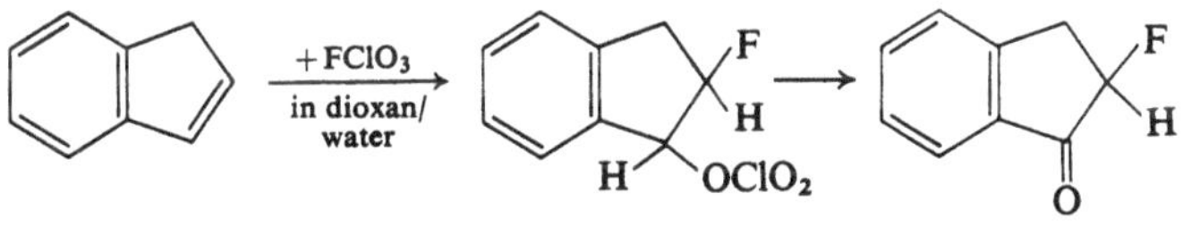

(4.15)

5 Reactions of molecular chlorine with unsaturated compounds

'The property of chlorine, to destroy offensive odours and to prevent putrefaction, gives to the chlorides of lime and soda a high value.' (Ure, *Dict. Arts*, vol. I, p. 781)

5.1 Introduction

Molecular chlorine is easily able to sustain free-radical chain reactions, especially when its dissociation into atoms is activated photochemically or by the use of free-radical initiators. It has already been noted that these reactions can lead to substitutions, as with neopentane, Me_4C; (2.2). With unsaturated compounds, however, addition usually prevails; benzene, for example, gives tetra- and hexa-chlorides, one of the geometrical isomers of the latter being the important insecticide gammexane: (5.1).

$$C_6H_6 + 3Cl_2 \longrightarrow C_6H_6Cl_6 \tag{5.1}$$

Thermal activation of free-radical additions is possible at relatively low temperatures even in the absence of light, provided that aprotic conditions are chosen and the olefinic compound is reasonably but not too strongly activated by substituents that facilitate the alternative competing heterolytic reaction.

In general, however, it is easier to achieve this alternative type of reaction, which is the subject of the present chapter. For addition in reasonably polar conditions, ethylene reacts very rapidly with chlorine at room temperature; when conventional methods are used for kinetic measurements, study only of less reactive olefinic compounds is possible, so compounds in which the double bond carries electron-withdrawing substituents must be chosen. In the aromatic series, however, electrophilic chlorination of benzene is very slow, and the kinetic study of compounds activated by electron-releasing substituents is more convenient.

Some important generalisations can be made concerning the heterolytic reactions of molecular chlorine with unsaturated compounds, be they olefinic, dienoid or aromatic in nature. The reactions are characterised by the prevalence of one very simple kinetic form; by strong response of reaction rate to change in structure or in solvent; and by reaction sequences which, though they involve carbocationic intermediates, appear at the same time to provide the departing halide ion with an important special role and to give complex product mixtures.

There have been a number of general reviews of electrophilic chlorination. For additions, reference may be made to accounts by Williams (1941), by de la Mare (1949), by de la Mare and Bolton (1966) and by Bolton (1973). Aromatic substitutions are dealt with in accounts by de la Mare and Ridd (1959), by Norman and Taylor (1965), by Fahey (1968) and by Taylor (1972). These provide sources for amplification of the factual material provided in this chapter but not otherwise specifically documented.

5.2   Kinetic forms: composition and nature of the transition state

Reactions of molecular chlorine with unsaturated compounds have been examined kinetically in a number of solvents, but in greatest detail in acetic acid and its mixtures with water. The simple kinetic form given in (5.2) has been observed uniformly.

$$-\mathrm{d[Unsaturated\ compound]}/\mathrm{d}t = k_2\mathrm{[Unsaturated\ compound][Cl_2]} \quad (5.2)$$

The effects of added electrolytes on the rate of reaction show that pre-equilibria such as those shown in (5.3)–(5.5) are unimportant.

$$Cl_2 \rightleftharpoons Cl^+ + Cl^- \quad (5.3)$$

$$Cl_2 + SOH \rightleftharpoons ClOS + HCl \quad (5.4)$$

$$Cl_2 + SOH \rightleftharpoons ClOSH^+ + Cl^- \quad (5.5)$$

The transition state for chlorination, therefore, whether substitution or addition is concerned, involves the unsaturated compound and the whole of the chlorine molecule. In general this transition state is more powerfully solvated, and hence must be more dipolar in character, than the starting materials, as follows from the fact that the rate of reaction is much increased by increase in the polarity of the solvent (e.g. by adding water to the solvent acetic acid). The reactions are usually irreversible, and substitution in aromatic compounds is not subject to a hydrogen–deuterium isotope effect of greater than unity, so proton loss has made no progress in the rate-determining step for this reaction. The transition state must therefore resemble a chlorocarbocation still associated with a chloride ion. The intermediate to which this transition state leads is represented as an ion-pair in structure (**5.1**), detailed consideration of its structure being deferred.

As far as additions are concerned, added chloride ions may sometimes contribute kinetically, adding the term given in (5.6) to the rate equation. The rate-determining stage under these circumstances may be the reaction of the intermediate (**5.1**) with chloride ions; the environmental effects of added salts make it difficult to establish the incursion of this pathway with certainty.

$$-\mathrm{d}[\,\text{Unsaturated compound}\,]/\mathrm{d}t = k_2[\,\text{Unsaturated compound}\,][\,\mathrm{Cl_2}\,][\,\mathrm{Cl^-}\,] \qquad (5.6)$$

$$\left[\begin{array}{c} \overset{+}{\underset{}{>}}\mathrm{C}\!-\!\mathrm{C}\!\overset{<}{\underset{\overset{|}{\mathrm{Cl}}}{}} \\ \mathrm{Cl^-} \end{array} \right]$$

(**5.1**)

Addition of chlorine to $\alpha\beta$-unsaturated carbonyl compounds (e.g. to cinnamaldehyde, Ph.CH:CH.CHO) can be complicated by acid-catalysis. It is probable (Cabaleiro *et al.*, 1968) that this mode of reaction involves 'protonation of the carbonyl group, followed by addition of a nucleophile to give a new substrate more reactive with molecular chlorine (e.g. Ph.CH:CH.CH(OH)OS, Ph.CH:CH.CH(OH)Cl, Ph.CH(Cl)CH:CH.OH), but the details are not yet known with certainty.

5.3 Effects of structure on reactivity

The rate of molecular chlorination of unsaturated compounds responds powerfully to change in structure; electron release to the reaction centre facilitates, and electron withdrawal from it retards the reaction. The kinetic results for situations in which the substituent is conjugated with the reaction centre can be expressed in linear free-energy terms through the use of values of the reaction constants, ρ, determined by plotting the relative rates of chlorination against the modified substituent constants σ^+ (see chapter 1). Some values of ρ determined in this way are given in table 5.1; their negative signs show that the reactions are favoured by electron release, a result which indicates that the transition state is carbocationoid.

Typical partial rate factors, which give the statistically corrected rates

TABLE 5.1 *Reaction constants, ρ, for some molecular chlorinations in acetic acid at 25 °C*

Substrate	ρ	Notes
$R.C_6H_5$	-11	Chlorination *para* to R; rates of chlorination of polycyclic aromatic compounds confirm the high value, which indicates a large response to change in structure.
$R.CH:CH.CO_2H$	-5.5	Correlation imperfect for $R = Ph$; see text.
$p\text{-}R.C_6H_4\ CH:CH.CO_2H$	-3.9	Correlation good.
$p\text{-}R.C_6H_4.C_6H_5$	$ca\ -3.3$	Chlorination *para* to $p\text{-}R.C_6H_4$; from measurements for $R = Me$ only.

of attack on individual positions of the substituted aromatic nucleus relative to the rate of substitution at a single position in benzene, are shown in structures (5.2)–(5.4), in which the values have been derived from the experimental results for relative rates of chlorination and for product proportions. The effects of the substituents in these compounds, and of some others, have been shown by study of di- and poly-substituted benzenes to be independent and additive in free energy of activation to a reasonably good approximation. Thus the effect of a *m*-methyl group in toluene, (5.2), is very nearly the same as that which would be derived from the rate of chlorination of *p*-methylacetanilide (structure (5.5); partial rate factor for a *m*-methyl group, $f_{Cl_2}^{m\text{-Me}} = 31 \times 10^5/6.1 \times 10^5 =$

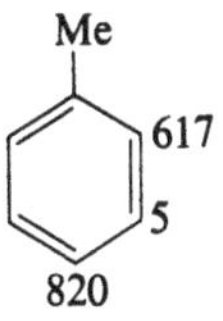

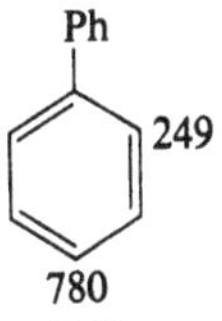

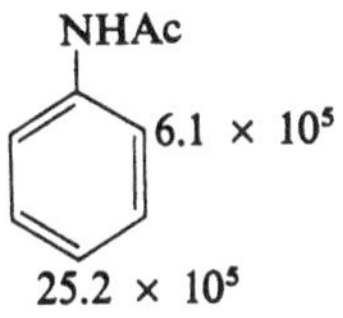

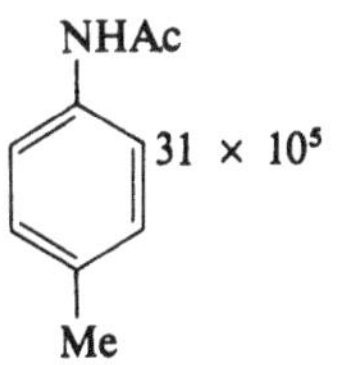

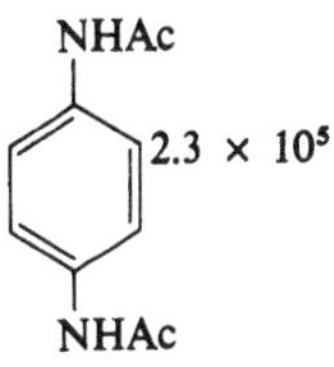

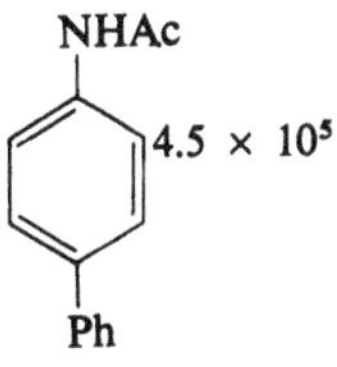

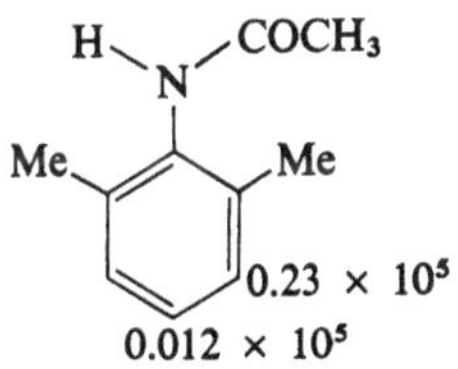

Partial rate factors for some molecular chlorinations

5.1). If it is assumed that this method can be applied more widely, estimates can be made of quantities which would be difficult, if not impossible, to determine experimentally. Thus from the experimental values for *p*-diacetamidobenzene, (**5.6**), and for *p*-acetamidobiphenyl, (**5.7**), the partial rate factors for the *m*-acetamido and *m*-phenyl substituents can be assessed as $f_{Cl_2}^{m\text{-NHAc}} = 0.4$ and $f_{Cl_2}^{m\text{-Ph}} = 0.7$. On this basis the chlorination of acetanilide should give only 2×10^{-5} per cent of *m*-chloroacetanilide, and that of biphenyl should give only 0.1 per cent of *m*-chlorobiphenyl.

This type of approach to reactivity and orientation in polysubstituted benzenes has had some measure of success, therefore; and some of the types of deviation from additivity of substituent effects are well recognised, understood and qualitatively predictable. Thus a substituent such as the acetamido-group can exert its maximum conjugative power only when it can become as near as possible to coplanar with the aromatic ring. The presence of a bulky *ortho*-group interferes with this geometric requirement. So, for example, 2,6-dimethylacetanilide is chlorinated more slowly than would otherwise be expected, and substitution is directed *ortho, para* to the methyl group rather than *para* to the acetamido group: (**5.8**). Other complications, however, can result from complexities in the product forming stages of chlorination, and will be discussed later.

For olefinic systems in which one terminus of the double bond is the favoured position of attack, the effects of substituents β to the point of attack can correctly be treated similarly, and such an assumption was implicit in the choice of substituent constants for derivation of the ρ-values given in table 5.1. It must be recognised, however, that the steric situation of a substituent in an olefinic system is different from that which prevails in an aromatic compound. This is illustrated for the phenyl group by the comparison of structures (**5.9**) and (**5.10**). In the planar

(**5.9**)   (**5.10**)

conformation shown, the two pairs of *ortho-*, *ortho'*-hydrogen atoms in biphenyl, (**5.9**), are each 180 pm apart, well within the normal touching distance (240 pm). So maximum conjugation is inhibited sterically; and this inhibition is reduced in the styrene system, (5.10), by the removal of one such non-bonding compression. For this reason, the effect of a single phenyl group on *para*-substitution in aromatic systems is proportionately less than on addition to olefinic compounds, and Hammett plots for the latter type of reaction show the phenyl group as considerably aberrant. With larger substituents in an appropriate blocking position, the activation of the double bond by an aryl group can, of course, be markedly diminished.

The values of the reaction constants, ρ, given in table 5.1 illustrate separately for attack on olefinic and on aromatic compounds that the effects of substituents on reactivity become attenuated when the substituent is removed further from the point of electrophilic attack. This principle does not apply, however, to comparison of the parent aromatic and olefinic compounds; substituent effects are at their largest for aromatic chlorination, and tend to be greater for *para-* than for *ortho-* substitution (cf. partial rate factors in structures (5.2), (5.3)). For attack on acyclic conjugated dienes, similar considerations apply; and it is accepted generally that attack occurs preferentially at the end, rather than at the centre, of the conjugated system. Thus piperylene (1,3-pentadiene, (5.11)) reacts with chlorine to give the 1,2- (and 1,4-) rather than the 3,4-adduct: (5.7). Quantitative studies of reactivity have not so far been attempted.

$$\text{Me.CH:CH.CH:CH}_2 + \text{Cl}_2 \longrightarrow \text{Me.CH:CH.CH(Cl).CH}_2\text{Cl} \quad (5.7)$$

(5.11)

Treatment of orientation in relation to the rate of addition to olefinic compounds becomes more complicated when significant attack can occur at both ends, or at the centre, of a substituted double bond. Two types of model might be envisaged. The first might be considered suitable for dealing with 'central' attack to give a cyclic structure having 'chloronium' character (e.g. the ion pair, (5.12)). Under these circumstances,

$$\left[\begin{array}{c} \text{R}_1 \\ \text{R}_2 \end{array}\!\!>\!\!\text{C}\!\!-\!\!\text{C}\!\!<\!\!\begin{array}{c}\text{R}_3\\\text{R}_4\end{array} \atop \begin{array}{c}\text{Cl}\\\text{Cl}^-\end{array}\right] \qquad \begin{array}{c}\text{R}_1\\\text{R}_2\end{array}\!\!>\!\!\text{C}\!\!\!\underset{\alpha}{=}\!\!\!\underset{\beta}{\text{C}}\!\!<\!\!\begin{array}{c}\text{R}_3\\\text{R}_4\end{array}$$

(5.12)   (5.13)

a single type of substituent constant (e.g. $\sigma^+_{p\text{-R}}$, $\sigma^+_{m\text{-R}}$ or some specified combination of these) together with a single ρ-value might be expected to give satisfactory correlation with the rate of addition. In the second, two types of partial rate factor might be recognised according to the relationship between the position of the substituent and the specific position of attack as indicated in structure (**5.13**). One of these (e.g. perhaps $\sigma^+_{p\text{-R}}$) would be appropriate to R_1 and R_2 activating C_β or to R_3 and R_4 activating C_α. The other (e.g. perhaps $\sigma^+_{m\text{-R}}$) would be appropriate to the activation of C_α by R_1 and R_2, or of C_β by R_3 and R_4. The total rate of reaction would then be represented by the sum of the derived partial rate factors. Allowance might have to be made on either model not only for steric hindrance of attack on particularly crowded positions but also for differences in strain energies of isomeric ground states.

Attempts have been made to use experimental results to distinguish between the two models. Fahey (1968), for example, has claimed that the relative rates of chlorination in aprotic solvents, $Me.CH:CH_2$ (1), $Me_2C:CH_2$, $Me.CH:CH.Me$ (50), $Me_2C:CHMe$ (10^4), $Me_2C:CMe_2$ (4×10^5) support the model based on a transition state analogous to structure (**5.12**) because successive methyl groups have approximately equal effects in enhancing the rate of reaction. In chapter 7, it will be seen however, that a similar, more extensive analysis of results of bromination suggests that in polar solvents the second model probably gives a better representation of the experiments.

5.4 Product-forming sequences: replacement at the attacked centre

5.4.1 Substitution in olefinic compounds. Although attention is usually focussed on additive chlorination of olefinic compounds, molecular chlorine often gives also products of substitution, even under conditions favourable to heterolytic reactions and under circumstances in which the products of addition are stable, so that an addition–elimination sequence is not responsible for the substitution. Displacements of this kind occurring without double-bond rearrangement are responsible for minor components of the product mixtures in the chlorination of isobutene (see also §§ 5.5 and 6.4.3), and can be exemplified also for derivatives of styrene: (5.8) (Fahey and Schubert, 1965). Since methyl cinnamate

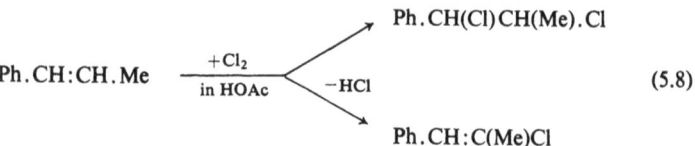

$$Ph.CH:CH.Me \xrightarrow[\text{in HOAc}]{+Cl_2} \begin{cases} Ph.CH(Cl)CH(Me).Cl \\ \\ -HCl \\ \\ Ph.CH:C(Me)Cl \end{cases} \qquad (5.8)$$

($Ph.CH:CH.CO_2Me$) is reported to give very little substitution under similar conditions (Cabaleiro *et al.*, 1967, 1968), it seems probable that substitution is somewhat favoured relative to addition by electron release to, rather than by electron withdrawal from, the reaction centre.

5.4.2 Substitution in aromatic compounds. For substitution in aromatic compounds, the major concern is with the directing power of substituents already present in the aromatic molecule. The results for molecular chlorination support the usual theories of aromatic substitution, which involve the assumption that the observed product proportions establish the relative rates of electrophilic attack on the various positions of the aromatic ring, and are not influenced by the relative rates of the product-forming steps. Some illustrative details of substituent effects in molecular chlorination have been given in §5.3, and show that both inductively and conjugatively electron-releasing substituents activate specifically and powerfully the *ortho-* and *para*-positions. One such substituent is the methyl group, which inductively activates the *meta*-position also. Inductively electron-withdrawing groups, on the other hand, deactivate positions *meta* to them, as is shown for the phenyl group and the aceta-mido group in the partial rate factors ($f_{m\text{-Ph}}$, 0.7; $f_{m\text{-NHAc}}$, 0.4) already mentioned.

5.4.3 Comparison of effects of substituents on *ortho-* and *para*-positions. The relative influence of substituents on attack at *ortho-* and *para*-positions in aromatic systems is controlled by a number of factors. First, *ortho-* and *para*-positions are intrinsically activated differentially, and the extent of this differentiation itself depends on the balance between conjugative and inductive effects. Secondly, a substitution having a higher ρ-value will differentiate more strongly between the partial rate factors for any two positions whose relative rates are controlled by energies, rather than entropies, of activation. Thirdly, effects of steric hindrance become significant for *ortho*-substitution when the entering reagent is large. Fourthly, mechanisms specific for *ortho*-substitution may sometimes make significant contributions.

For molecular chlorination, the combination of factors has the consequence that, as in nitration, activated mono-substituted benzenes usually give mixtures of *ortho-* and *para*-derivatives. Steric hindrance to entry of the attacking chlorine molecule probably becomes significant only for groups as large as or larger than the methyl group. For the t-butyl group, it has been calculated (de la Mare, 1958; de la Mare and Ridd, 1959) that it amounts to about 4.5 kJ mol^{-1}, with the consequence that t-butylbenzene gives only about 22 per cent of the *o*-chloro-derivative.

5.4.4   Replacement of groups other than hydrogen. Chlorination with replacement of groups other than hydrogen is known both for olefinic and for aromatic compounds. Cabaleiro, Johnson, Swedlund and Williams (1968) have reported that chlorodecarboxylation of the cinnamate ion accompanies addition in acetic acid: (5.9). Displacement of the

$$\text{Ph.CH:CH.CO}_2^- \xrightarrow[\text{in HOAc}]{+\text{Cl}_2} \text{Ph.CH:CH.Cl} + \text{CO}_2 + \text{Cl}^- \quad (5.9)$$

t-butyl group in the course of the chlorination of t-butylbenzene, (5.10), has also been reported as a very minor, but definitely identified, component of the reaction path. Calculation of a partial rate factor for *ipso*-substitution for this reaction gives a value of about 1. The conventional

$$\text{t-Bu.C}_6\text{H}_5 \xrightarrow[\text{in 99\% HOAc}]{+\text{Cl}_2} \text{Cl.C}_6\text{H}_5 \; (0.13\%) + \text{Cl}^- + \text{t-Bu}^+ \quad (5.10)$$

interpretation of this value would be that the rate of attack by chlorine on the *ipso*-position in t-butylbenzene is about equal to that of attack on a single position in benzene. A number of unproved assumptions have to be made in making this deduction, however; among them are that the general mechanism is that of the normal chlorine substitution and that the first stage of de-t-butylation is not reversible.

Chlorodesilylation of trimethylsilylbenzene, (5.11), has been investigated kinetically. The reaction follows the expected kinetic form, (5.12),

$$\text{Ph.SiMe}_3 \xrightarrow[\text{in HOAc}]{+\text{Cl}_2} \text{Ph.Cl} + \text{Cl.SiMe}_3 \quad (5.11)$$

$$-\text{d}[\text{Cl}_2]\text{d}t = k_2[\text{Ph.SiMe}_3][\text{Cl}_2] \quad (5.12)$$

and is presumed to be a conventional aromatic replacement involving a carbocationic transition state and intermediate, though details of the product-forming steps have not been investigated.

Replacement in the course of aromatic chlorination is known also for other substituents: cf. (5.13) and (5.14); (Datta and Bhoumik, 1921; Elion, 1923, 1925). Though these reactions have not been investigated mechanistically, they almost certainly involve electrophilic attack to form an intermediate which then loses the displaced substituent.

$$p\text{-Br.C}_6\text{H}_4.\text{SO}_3\text{H} \xrightarrow{+\text{Cl}_2+\text{H}_2\text{O}} p\text{-Br.C}_6\text{H}_4.\text{Cl} + \text{H}_2\text{SO}_4 + \text{HCl} \quad (5.13)$$

$$p\text{-H}_2\text{N.C}_6\text{H}_4.\text{CO}_2\text{H} \xrightarrow{+\text{Cl}_2} p\text{-H}_2\text{N.C}_6\text{H}_4.\text{Cl} + \text{CO}_2 + \text{HCl} \quad (5.14)$$

5.5   Product-forming sequences: replacement with rearrangement

Chlorination with accompanying double-bond rearrangement is well known both in olefinic and in aromatic systems. Though the scope of these reactions has not been explored systematically, general experience seems to suggest that they are most commonly encountered in highly activated systems. The main product of reaction of isobutene and chlorine in the liquid phase at low temperatures is 2-methylallyl chloride, formed by attachment of chlorine to the unsaturated carbon atom, (5.15), as was shown by experiments using ^{14}C-labelled isobutene. The intermediate can be assumed to have the bridged structure (**5.14**); evidence supporting this view is given in §6.4.3.

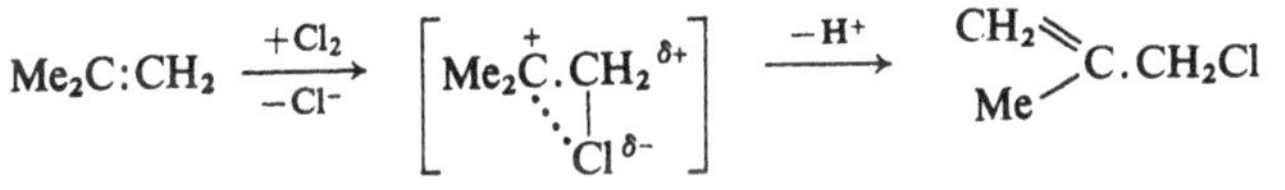

$$Me_2C:CH_2 \xrightarrow[-Cl^-]{+Cl_2} \left[\overset{+}{Me_2C} . CH_2^{\delta+} \underset{\overset{|}{Cl^{\delta-}}}{} \right] \xrightarrow{-H^+} \underset{Me}{\overset{CH_2}{\diagdown}} C . CH_2Cl \quad (5.15)$$

$$(\textbf{5.14})$$

The smooth and rapid side-chain chlorination of hexamethylbenzene and related compounds probably involves a similar sequence followed by an allylic rearrangement: (5.16). Here the rate of disappearance of

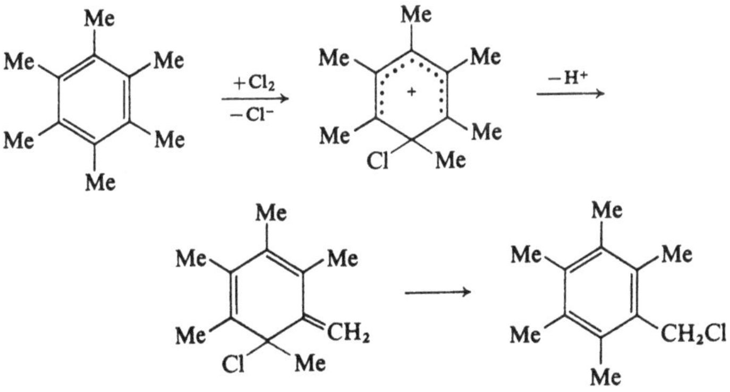

$$(5.16)$$

chlorine follows the usual kinetic form, and is increased as expected by electron release to the reaction centre. Cross-conjugated trienes may be implicated as intermediates also, as has been emphasised particularly for the corresponding reaction of 1,2,3,5-tetramethylbenzene (Baciocchi, Mandolini and Patara, 1975).

In these examples, proton loss occurs from a C–H bond attached to a carbon atom at which a cationic charge has developed. Similar reactions involving O–H bonds are well authenticated. Thus the chlorination,

(5.17), of 2,4-dichloro-1-naphthol gives 2,4,4-trichloro-1-oxo-1,4-dihydronaphthalene. Reasonably stable dienones are most readily

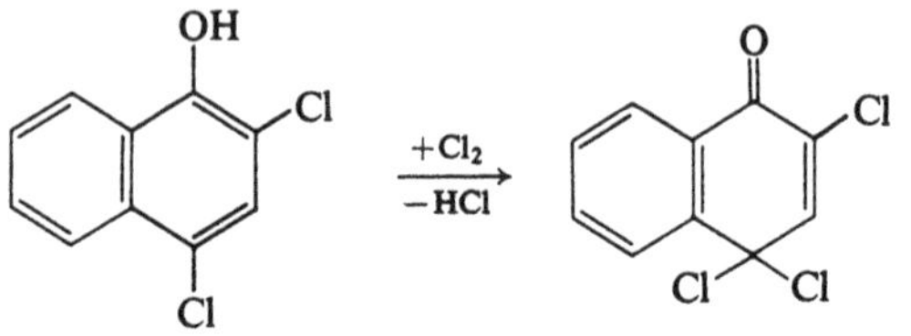

(5.17)

obtained by such reactions when there is no unsubstituted position in the aromatic ring activated by the electron-releasing group, so that there is no hydrogen available for subsequent prototropic rearrangement. It is not yet known whether chlorination of phenols generally involves displacement with rearrangement.

The corresponding processes involving N–H bonds must also be possible; an example (Bell, 1953) is given in (5.18). Mechanistic details

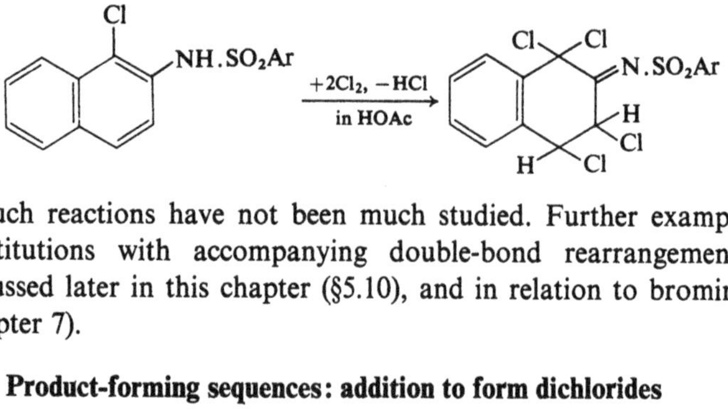

(5.18)

of such reactions have not been much studied. Further examples of substitutions with accompanying double-bond rearrangement are discussed later in this chapter (§5.10), and in relation to bromination (chapter 7).

5.6 Product-forming sequences: addition to form dichlorides

5.6.1 *anti*-Addition; addition to alkyl-substituted ethylenes, to deactivated olefinic systems, and to simple cyclo-alkenes. The product-forming steps leading to addition to simple olefinic systems, including those bearing electron-withdrawing substituents, normally result in *anti*-addition. It has usually been assumed that this can best be explained by assuming that the reaction involves an intermediate in which the chlorine partly or completely bridges the double bond, as in structures (5.15), (5.16), (5.17) or (5.18). Chloride ion then attacks this intermediate to give the product of *anti*-addition. The entering chloride ion can be derived either from the attacking chlorine or from the environment; when the reaction is carried out in an aqueous medium, the yield of dichloride is usually increased by adding chloride ions to the medium.

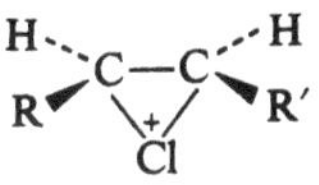

(5.15) 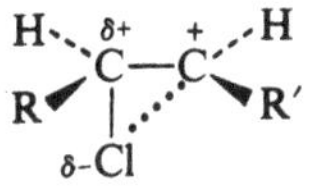 (5.16) 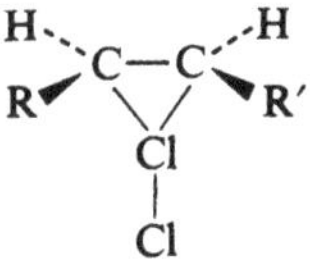 (5.17) 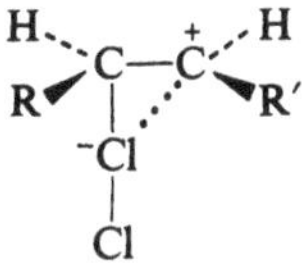 (5.18)

This stereochemistry prevails even for such a substrate as *cis*-di-t-butylethylene, which is considerably strained internally by non-bonding interactions between the t-butyl groups. Bridging by the attacking chlorine must, therefore, be energetically quite large if it provides the correct explanation of the stereochemistry of addition. An alternative possible way in which *anti*-addition could be determined involves attack of chlorine and the nucleophile in so nearly synchronous a fashion that the opening of the bridged ion and its internal rotation, (5.19), does not have time to occur. This interpretation is perhaps more attractive, because of the known weakness of neighbouring-group interaction involving chlorine; unfortunately the chlorination of di-t-butylethylene is too rapid for convenient kinetic study.

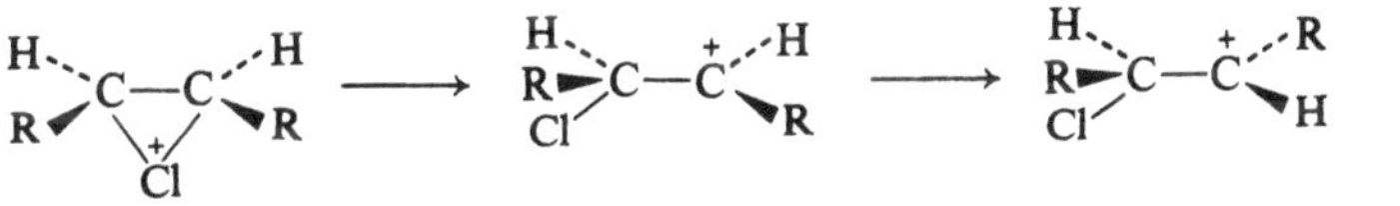

(5.19)

Addition to cyclo-alkenes also proceeds with *anti*-stereochemistry; and the fact that cholest-2-ene (partial formula, structure (5.19)) gave mainly the diaxial dichloride (partial formula, structure (5.20)) has been taken as indicating that attack occurs in preference approximately at right angles to (rather than in) the plane of the double bond.

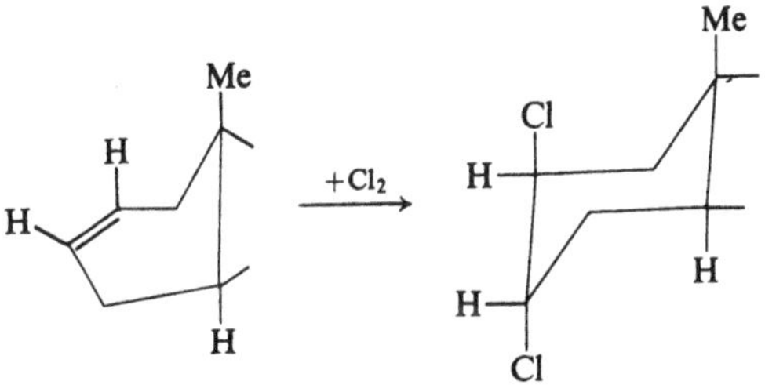

(5.20)

Cholest-2-ene $2\beta,3\alpha$-Dichlorocholestane
(5.19)   (5.20)

5.6.2 *syn*-Addition; addition to aryl-substituted cyclic systems, including aromatic compounds.
For molecular chlorine, several pathways are available leading to *syn*-addition, which therefore can be exemplified

just as widely as *trans*-addition can. An important example was provided by Cristol, Stermitz and Ramey (1956) who showed that acenaphthylene reacts with chlorine in chloroform with iodine as a catalyst to give, (5.21), predominantly the *cis*-adduct.

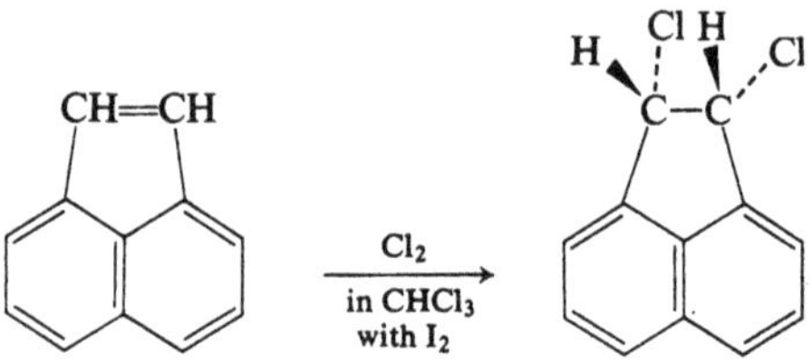

(5.21)

The similar addition to phenanthrene has been studied kinetically in acetic acid and in other solvents, and has been compared with the corresponding chlorination of naphthalene. The latter reaction is more complicated, since heterolytic addition to give the *cis*-dichloride is

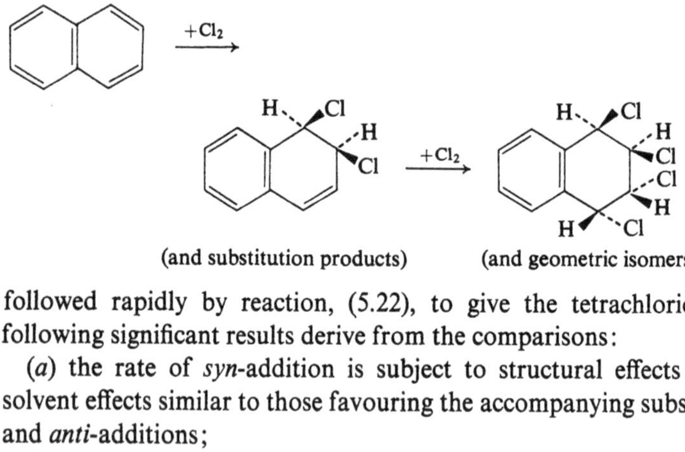

(and substitution products)   (and geometric isomers)

(5.22)

followed rapidly by reaction, (5.22), to give the tetrachloride. The following significant results derive from the comparisons:

(*a*) the rate of *syn*-addition is subject to structural effects and to solvent effects similar to those favouring the accompanying substitution and *anti*-additions;

(*b*) added chloride ion only slightly diverts the addition towards the formation of the *trans*-dichloride.

This mode of *syn*-addition appears, therefore, to involve a carbocationoid intermediate (e.g. (**5.21**) or (**5.22**)) formed early in the reaction path; only at a later stage is it possible to divert addition by nucleophilic capture, and this process probably involves a second intermediate. The resemblance of (**5.21**) to the intermediate considered to be involved in *cis*-fluorination (chapter 4) is clear; but for chlorination the development of carbocationic charge has been established, and the possibility that the covalent Cl–Cl bond is partly retained in the intermediate, (**5.22**), needs serious consideration.

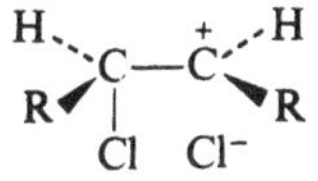

(5.21) 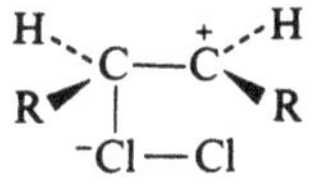   (5.22)

syn-Addition determined through an intermediate such as is shown in (5.21) and (5.22) is not confined to double bonds concerned in cyclic aromatic resonance; thus some *syn*-addition occurs in the chlorination of *trans*-1,2-dichloro-1,2-dihydronaphthalene, (5.23); Burton *et al.* (1974). Stabilisation of the cationic charge by conjugation seems to be one of the subtle factors determining the importance of this mode of reaction, which has also been observed for addition to D-glucal triacetate, (5.23) (Igarashi *et al.*, 1968), probably for a similar reason.

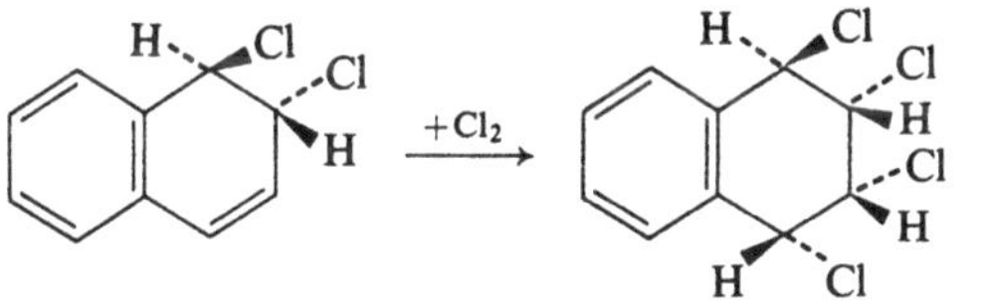

(5.23)

and other geometric isomers

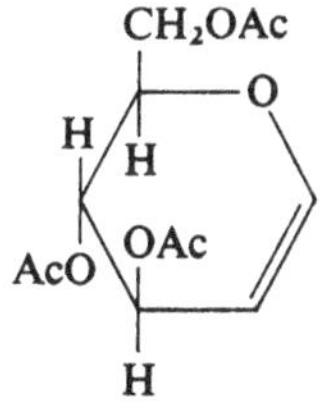

(5.23)

5.6.3 *syn*-Addition; addition to aryl-substituted acyclic systems.

syn-Addition has been observed also, accompanied by *anti*-addition, for the reaction of chlorine in acetic acid or trifluoroacetic acid with aryl-substituted olefinic compounds, including the 1-phenylprop-1-enes (Ph.CH:CH.Me), a number of substituted *cis*- and *trans*-cinnamic acids and their derivatives (R.C$_6$H$_4$.CH:CH.CO$_2$R′), and some aryl-substituted $\alpha\beta$-unsaturated ketones (Ar.CH:CH.COAr′). The 'direct' route just discussed (§5.6.2) provides one way in which these results could be interpreted. Another involves conformational change through internal rotation of an open carbonium ion, as in (5.19), prior to capture of the nucleophile; or alternatively, opening of the bridge followed by internal or external migration of chloride ion from one face of the original double-bond system to the other. It is probable that both 'direct' and 'open' pathways occur in chlorination of these compounds; and furthermore, the 'open' pathway may be forced on the system by the introduction of sufficiently large steric interactions, since *cis-* and *trans*-1-t-butyl-2-phenylethylenes (Ph.CH:CH.CMe$_3$) are chlorinated to give the same mixture of *erythro-* and *threo*-dichlorides (Abrahams and Monasterios, 1973). More characteristically, however, these chlorinations do not involve a carbocationic intermediate in conformational equilibrium, since normally *cis-* and *trans*-isomers give different mixtures of isomeric products.

5.6.4 *syn*-Addition; competition for the carbocationic centre by other neighbouring groups. Another route by which *syn*-addition can be determined is illustrated by the sequence shown in (5.24). This mode of addition can, under some conditions, become nearly the exclusive pathway taken in the chlorination of derivatives of maleic and fumaric acid;

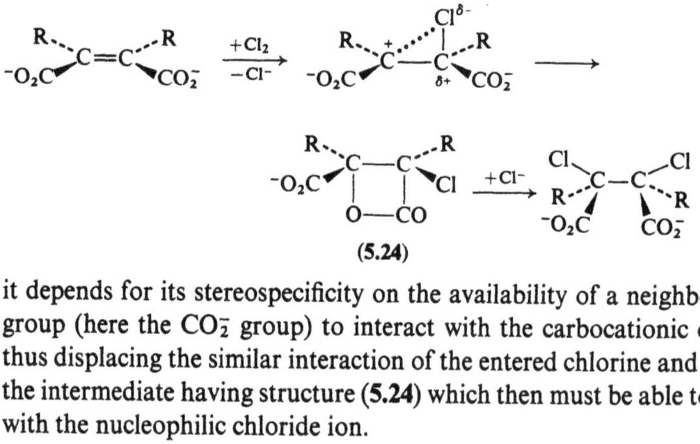

(5.24)

it depends for its stereospecificity on the availability of a neighbouring group (here the CO$_2^-$ group) to interact with the carbocationic centre, thus displacing the similar interaction of the entered chlorine and giving the intermediate having structure (**5.24**) which then must be able to react with the nucleophilic chloride ion.

5.6.5 *syn*-Addition determined by solvent interactions. Yet another way in which *syn*-addition of chlorine could be promoted involves the intervention of solvent in a sequence such as that shown in fig. 5.1. The essential feature of this reaction path is the fact that the interaction of the carbocationic intermediate with solvent gives a product (**5.25**) which, like the intermediate having structure (**5.24**), does not remain as the stable product, but instead reacts with chloride ion to give the product of *syn*-addition. It is unlikely that such a reaction path would result in exclusive

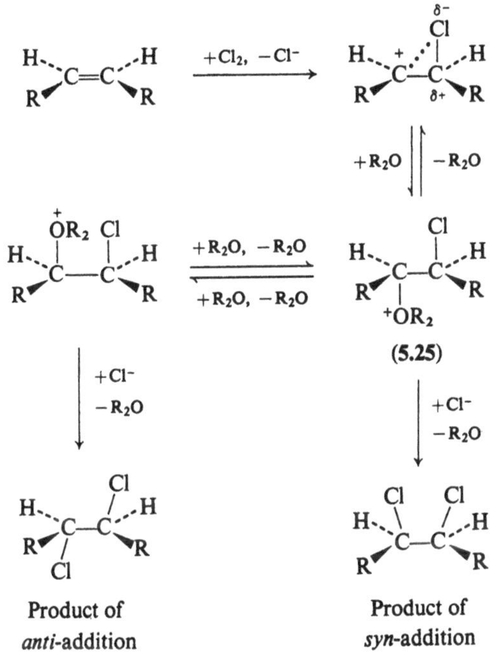

Fig. 5.1. *syn*-Addition determined by reversible interaction of a carbocationic intermediate with the solvent.

syn-addition, since internal rotation, or exchange with solvent giving a further formal inversion of configuration at the centre to be attacked finally by the chloride ion, would ensure that *syn*- and *anti*-addition would occur together. Similar reaction paths have been considered in relation to the stereochemistry of solvolysis of alkyl halides; and an example which may be of this kind is the reportedly non-stereospecific addition of chlorine to cyclohexene in dimethylformamide as solvent (De Roocker and de Radzitzky, 1970).

5.7   Product-forming sequences: addition with accompanying incorporation of nucleophiles

5.7.1   Attack by an external nucleophile.
It will already have become apparent that one of the well-authenticated fates for a carbocationic intermediate is its capture by a nucleophile. This nucleophile may be an anion; or an anion provided by a solvent molecule with loss of a proton; or a neutral molecule, which may be a molecule of solvent. Equations (5.25) and (5.26) provide specific examples from the chemistry of aliphatic compounds; (5.27) comes from aromatic chemistry.

$$CH_2\!:\!CH_2 \xrightarrow[-HCl]{+Cl_2,\ +H_2O} Cl.CH_2.CH_2.OH \qquad (5.25)$$

$$Ph.CH\!:\!CH_2 \xrightarrow[-Cl^-]{+Cl_2,\ +O:CH.NMe_2} \begin{array}{l} Ph.CH.CH_2.Cl \\ \quad | \\ \overset{+}{O}.CH\!:\!NMe_2 \end{array} \qquad (5.26)$$

$$\xrightarrow[-Cl^-]{+Cl_2,\ +OAc^-} \qquad (5.27)$$

In principle, the nucleophile can participate in the rate-determining step of the reaction, either by evoking a synchronous termolecular process or by attacking the carbocationic intermediate formed in pre-equilibrium; it can also be captured subsequent to the rate-determining formation of the carbocationic intermediate. In practice, it is difficult to distinguish amongst these possibilities, and it is likely that more than one of them can occur. Very little is known concerning the circumstances in which one or other pathway may prevail; too little to enable prediction concerning the extent to which one or other of these modes of addition will compete with the accompanying formation of dichlorides, or with substitution. The best generalisation available at the present time is that interceptions by added anions or by solvent are often significant, but even then are often less effective than would have been expected. The most frequent stereochemical consequence of this mode of reaction is *anti*-addition. This is anticipated when interaction between the entering chlorine and the carbocationic centre holds the configuration, or when a termolecular process prevails; *syn*-addition is possible, however, when neighbouring-group interaction is relatively ineffective, or with certain other of the sequences which can lead to *syn*-dichlorides (§5.6).

5.7.2 Attack by an internal nucleophile. Reference to (5.24) enables the inference to be made that the diverting nucleophile can be provided internally; and indeed, β-lactones such as that formulated in structure **(5.24)** can be isolated from the chlorination of sodium dimethylmaleate or sodium dimethylfumarate. Other neighbouring groups can provide the internal nucleophilic centre, and rings of other sizes can be formed, as is illustrated later for other halogenating species (chapter 7).

5.7.3 Orientation; Markownikoff and anti-Markownikoff addition. In (5.26) it was illustrated that addition accompanied by incorporation of a nucleophile is often subject to orientational control by the substituent or substituents attached to the olefinic link, so becoming regiospecific or regioselective in character. In the example mentioned, this orientational specificity arises because the phenyl group in styrene activates the β-olefinic carbon atom by the process of electron release indicated by the arrows in structure (**5.26**), and hence electrophilic chlorine attacks specifically the terminal position. This conjugative influence is one of polarisability rather than one of polarisation; the similar structure (**5.27**) determines the orientation taken in the chlorination of vinyl chloride, (5.28), despite the large inductive effect in the reverse direction. For

$$\overset{\curvearrowleft}{Ph}.\overset{\curvearrowleft}{CH}:\overset{\curvearrowleft}{CH_2} \qquad\qquad \overset{\curvearrowleft}{Cl}.\overset{\curvearrowleft}{CH}:\overset{\curvearrowleft}{CH_2}$$

$$\textbf{(5.26)} \qquad\qquad\qquad\qquad \textbf{(5.27)}$$

$$CH_2:CH.Cl \xrightarrow[-HCl]{+Cl_2,\ +H_2O} Cl.CH_2.CH(OH).Cl \xrightarrow{-HCl} Cl.CH_2.CHO \qquad (5.28)$$

cases in which the electronic influence of the directing substituent is less pronounced, however, orientation is less regiospecific; and in some cases the results for chlorination contrast with those for addition of hydrogen chloride: (5.29) and (5.30). It seems natural to associate this orientational

$$CH_2:CH.CH_2Cl \xrightarrow{+HCl} H.CH_2.CH(Cl).CH_2Cl \qquad (5.29)$$

$$CH_2:CH.CH_2Cl \xrightarrow[-HCl]{+Cl_2,\ +H_2O} \begin{Bmatrix} Cl.CH_2.CH(OH).CH_2Cl\ (30\%) \\ HO.CH_2.CH(Cl).CH_2Cl\ (70\%) \end{Bmatrix} \quad (5.30)$$

difference with a difference in the importance of neighbouring-group interaction, which is small or non-existent for neighbouring hydrogen in the carbocationic intermediate (**5.28**) involved in hydrochlorination, but becomes significant in the corresponding intermediate (**5.29**) involved in chlorination.

$$\overset{\displaystyle H}{\underset{\displaystyle CH_2.\overset{+}{CH}.CH_2Cl}{|}} \qquad\qquad \overset{\displaystyle \delta-Cl}{\underset{\displaystyle \delta+CH_2.\overset{+}{CH}.CH_2Cl}{\overset{\cdots}{|}}}$$

$$\textbf{(5.28)} \qquad\qquad\qquad \textbf{(5.29)}$$

The significance of results relating to regiospecificity in addition are discussed in more detail in chapters 6 and 7, when additions initiated by other halogenating reagents are considered. Comment is necessary at this point, however, on the use of the terms 'Markownikoff' and 'anti-Markownikoff' in describing orientations of addition. The Markownikoff rule was originally formulated to describe the direction of addition of hydrogen halides to olefinic hydrocarbons in which the carbon atoms of the double bond held different numbers of hydrogen atoms. Now that a reasonable theoretical basis for interpretation of such orientation can be derived from the electronic theories of organic chemistry, it is possible to regard this rule as extensible on the basis of these theories: first, from a reaction such as that of (5.31) (Markownikoff addition of H–Cl) to the corresponding addition of water or generally of reagents H–X, (5.32); secondly, from these cases to those involving addition to olefinic compounds bearing other substituents (e.g. (5.29)); thirdly, to reactions initiated by other electrophiles (e.g. (5.26) and (5.28), which represent Markownikoff addition of Cl–X, and (5.30), which represents predominant anti-Markownikoff addition of Cl–X); and fourthly, to additions to olefinic compounds of the type R.CH:CH.R' or RR'C: CR''R'' (e.g. (5.33), which illustrates Markownikoff addition of ClOAc to β-nitrostyrene). Care is needed, however, in these extended uses. Care

$$\text{Me.CH:CH}_2 \xrightarrow{+\,\text{HCl}} \text{Me.CH(Cl).CH}_3 \qquad (5.31)$$

$$\text{Me.CH:CH}_2 \xrightarrow{+\,\text{H.OH}} \text{Me.CH(OH).CH}_3 \qquad (5.32)$$

$$\text{Ph.CH:CH.NO}_2 \xrightarrow{+\,\text{ClOAc}} \text{Ph.CH(OAc).CH(Cl).NO}_2 \qquad (5.33)$$

is also needed to avoid confusion between anti-Markownikoff addition (the reverse of Markownikoff addition) and *anti*-Markownikoff addition (Markownikoff addition with *anti*-stereochemistry).

5.8 Product-forming sequences: addition with double-bond rearrangement

A conjugated unsaturated compound such as 1,3-butadiene would be expected to react terminally with electrophilic chlorine for reasons of

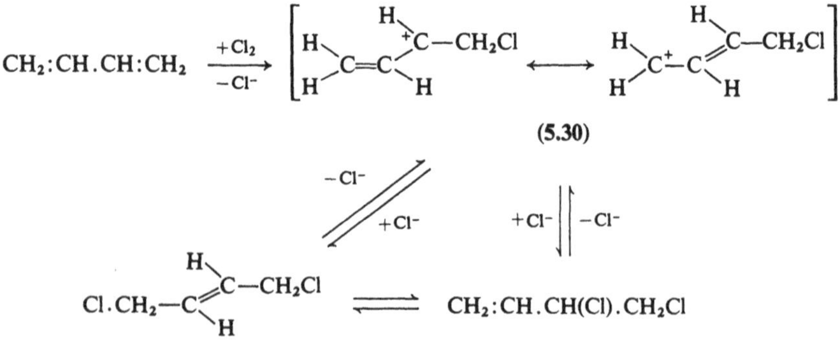

(5.30)

(5.31)   **(5.32)**

Fig. 5.2. Simplified representation of the reaction of 1,3-butadiene with chlorine.

polarisability, as is illustrated for the analogous case of styrene by the arrows in structure (5.26). In the resulting carbocationoid intermediate (simplified representation, structure (5.30), fig. 5.2), the cationic charge is delocalised through the contribution of the canonical structures shown. Further reaction with a nucleophile then can occur either at the 2- or at the 4-position; in the latter case, the product, (5.31), has been formed by 1,4-addition with rearrangement, and could be produced reversibly from the 1,2-adduct, (5.32), through the same carbocationic intermediate, or by intramolecular rearrangement. Figure 5.2 shows the inter-relationship of these possible reactions. In this case, the 1,2-adduct, (5.32), is thermodynamically less stable than the isomeric 1,4-adduct, (5.31), but the product of chlorination contains about 70 per cent of the former. The reaction is therefore kinetically controlled, and the passage from the intermediate to the less thermodynamically stable product is faster than the corresponding reaction to give the thermodynamically more stable product.

The 1,4-adduct has exclusively the *trans*-configuration. Steric influences would be expected to favour the formation of the intermediate in the conformation shown in structure (5.30) rather than that shown in the isomeric form, (5.33). It seems clear that no cyclic interaction involving the entering chlorine and both termini of the unsaturated system (hypothetical structure (5.34) or related ring structures) has any importance, since the *cis*-1,4-dichloride (5.35) is not formed. The energy relationships for the conversion of the intermediate into products are shown approximately in fig. 5.3.

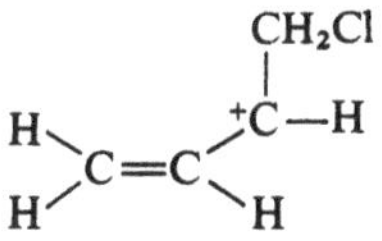

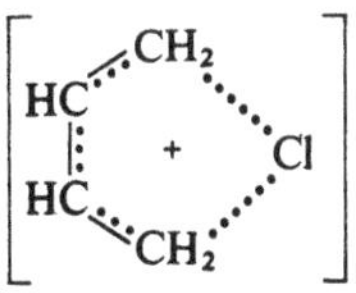

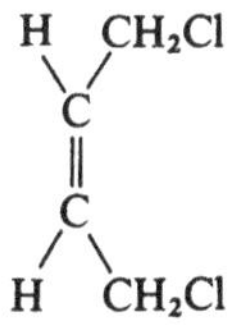

A number of other additions of chlorine to conjugated dienes proceed similarly to give mixtures of isomers. Sometimes the predominant product is that of *vic*-addition, as with 1-phenylbutadiene, (5.34); and sometimes that of 1,4-addition, as with isoprene, (5.35). For most of

$$Ph.CH:CH.CH:CH_2 + Cl_2 \longrightarrow Ph.CH:CH.CH(Cl).CH_2Cl \quad (5.34)$$

$$CH_2:C(Me).CH:CH_2 + Cl_2 \longrightarrow Cl.CH_2.C(Me):CH.CH_2Cl \quad (5.35)$$

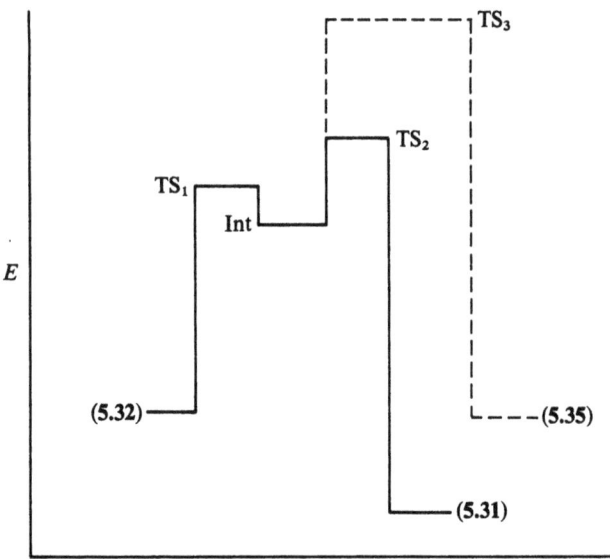

Fig. 5.3. Schematic energy diagram for the intermediate and products in the chlorination of 1,3-butadiene.
TS = transition state; Int = intermediate carbocation (5.30); (5.31) = *trans*-1,4-adduct (formed slowly, via TS₂); (5.35) = *cis*-1,4-adduct (not formed; TS₃ very high); (5.32) = 1,2-adduct (formed rapidly, via TS₁).

these cases, the question of whether the product has been formed under thermodynamic or kinetic control has not yet been answered with certainty.

In principle, completion of addition with double-bond rearrangement is possible with aromatic systems. The best-known case involves addition to anthracene, as in (5.36). Naphthalene, on the other hand, apparently gives substantially or exclusively 1,2-addition, (5.22): for, if 1,4-addition

$$\text{anthracene} + Cl_2 \longrightarrow \text{product} \qquad (5.36)$$

were important here, added chloride needed to effect this reaction and not available internally for steric reasons would increase the proportion of the pathway leading to the major tetrachloride, and this does not happen. Here the 1,2-adduct is expected to be the thermodynamically more stable product, since in it the resonance-stabilisation of a double bond conjugated with a benzene ring is maintained. The energy diagram

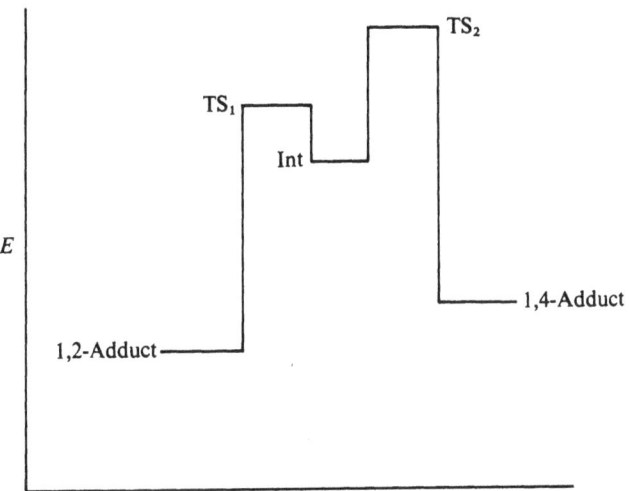

Fig. 5.4. Schematic energy diagram for the intermediate and main products of addition of chlorine to naphthalene.

shown in fig. 5.4 shows the relationship here between the paths leading from the postulated intermediate to the possible products; it contrasts with the case illustrated in fig. 5.3, in that the thermodynamically more stable product is the more rapidly formed.

There is no reason to doubt that fig. 5.2 represents the general nature of these conjugate additions, associated as they are with 1,2-additions. The mesomeric carbocationoid intermediate may, of course, have a more complicated structure than that shown for simplicity in the figure. The reason for the frequent preponderance of 1,2-addition is not so clear, however. It is not correct to associate this (in the sense of effect and cause) with the formation of the thermodynamically less stable isomer, though the generalisation that carbocationic intermediates often give more than would be expected of the thermodynamically less stable product is still a proper one. Perhaps the most useful generalisation is that in addition to 1,3-dienes and related compounds there often seems to be a kinetically easy route to 1,2-addition, irrespective of product stability; but that anionotropically related products are often formed also, even under kinetic control. It seems clear that any interaction between the entering reagent and the adjacent carbocationic centre, whether it is of the form that leads to *syn*-addition, (**5.21**) or (**5.22**), or is any of the bridged forms leading to *anti*-addition, (**5.15**)–(**5.18**), would help to localise the developing charge near to the entering electrophile, and so would favour 1,2-addition.

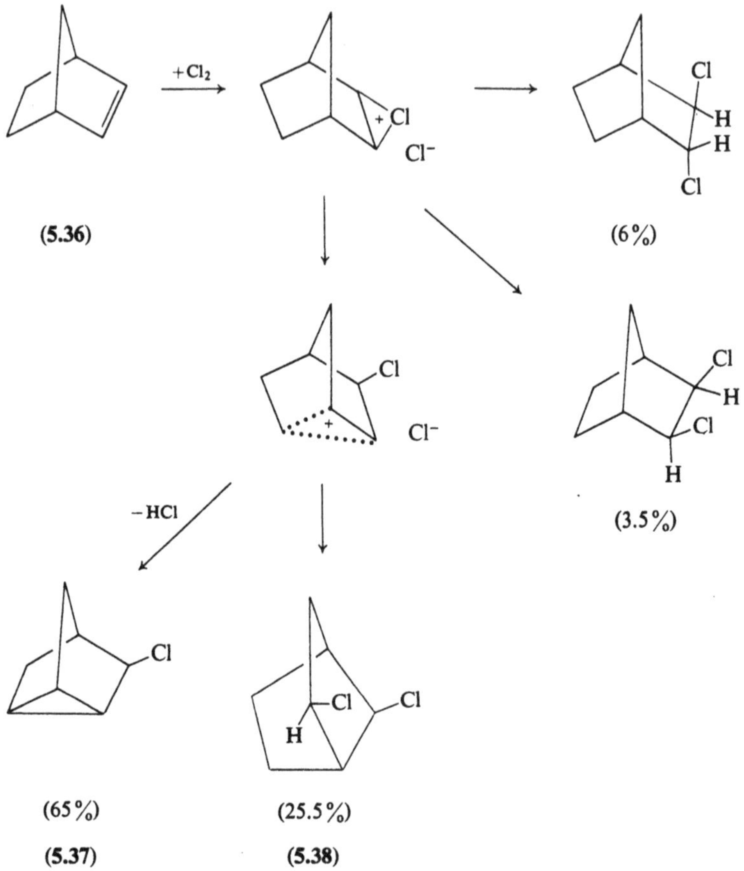

Fig. 5.5. Probable reaction path in the chlorination of norbornene.

5.9 Product-forming sequences: reactions through rearranged carbocationic intermediates

Carbonium ionic rearrangements accompanying chlorination are most prominent for cyclic olefinic structures, where the driving force enabling rearrangement to compete effectively with bridging by neighbouring chlorine is probably often enhanced by internal steric strain. Of the many well-documented examples, addition to norbornene (structure (**5.36**); fig. 5.5) is quoted here. When homolytic chlorination is suppressed, the main products are those of competition between internal trapping of the carbocationic centre, (**5.37**), and capture of this centre by nucleophilic chloride ion to give the rearranged dichloride, (**5.38**). The presumed reaction path is summarised in fig. 5.5.

Rearrangements involving migration of groups other than alkyl groups are also known, but have been investigated more fully for additions initiated by reagents other than molecular chlorine, so discussion is deferred until chapter 6.

5.10   Product-forming sequences: further reactions of the primary products

A feature of the chemistry of reactions of some unsaturated compounds with electrophiles is that the first-formed products may undergo further reaction with the electrophile to give a new set of products. Normally, since chlorine is an electron-withdrawing substituent, its attachment to an unsaturated system has a deactivating effect which is unfavourable to further reactions with the electrophile, so that the first reaction prevails provided that the organic substrate is present in sufficient concentration. Major exceptions occur, however. One of these concerns aromatic systems in which the first stage of addition leaves a double bond highly activated for substitution or addition. An example has already been given in the description, (5.22), of the course taken in the additive chlorination of naphthalene. Here the 1,2-adduct captures a further chlorine molecule so rapidly that it does not survive in the reaction mixture even when naphthalene is in considerable excess.

The failure to observe such a reaction can be used to make mechanistic deductions. Thus when the corresponding dichlorides are formed in the course of chlorination of a derivative of benzene, it will be expected that the reaction will proceed to give tetra- and hexa-chlorides: (5.1) and (5.37). The absence of such products in many substitutive chlorinations

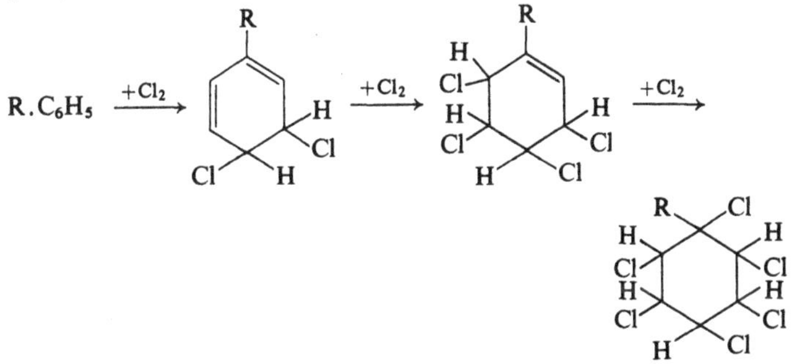

makes it probable, therefore, that the substitution products do not derive from 1,2-adducts; and conversely, when such adducts are found in significant amounts in the products of such a chlorination, (5.38), the competing importance of addition as a reaction path becomes obvious.

$$C_6H_5 . Ph \xrightarrow[\text{in HOAc}]{+2Cl_2}$$

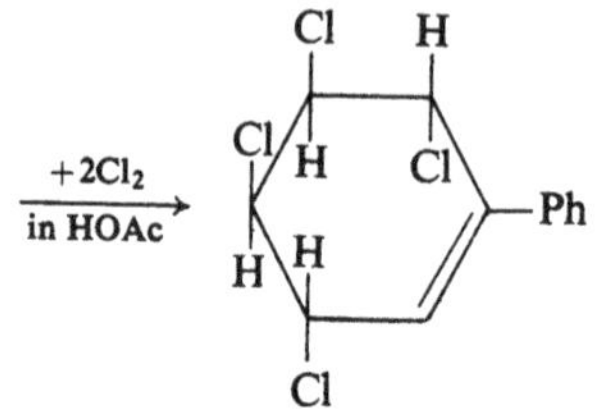

(5.38)

(9% of the hydrocarbon
consumed)

In fig. 5.6 is presented a simplified representation of the reaction pathway established for the chlorination of 1-methylnaphthalene. This chlorination follows a rather more involved course than that taken in the corresponding chlorination of 2-methylnaphthalene, which, like naphthalene, also gives mixtures of products of substitution and addition. From 1-methylnaphthalene, there can be recognised the products:

(*a*) of normal substitution (structure (**5.39**) and its isomers);

(*b*) of addition of chlorine, followed by further addition of chlorine, (**5.40**);

(*c*) of addition of chlorine, followed by addition of chlorine accompanied by interception with acetate ions, it not being excluded that these two reactions can occur in the reverse order, (**5.41**);

(*d*) of addition of chlorine, followed by substitution with rearrangement, followed by addition of chlorine, (**5.42**).

There are other situations in which the first stage of chlorination can produce an intermediate product more reactive than the starting material. Reactions involving chlorodeacylations with rearrangement give, after prototropic rearrangement, chlorophenols more reactive by many powers of ten than the original aryl acetate from which they were derived. Polysubstitution is expected, and is often observed, to the extent to which chlorodeacylation occurs as a component of the chlorination. The reaction of acetoxybenzene with chlorine, for example, is reported as giving a small proportion of polychlorophenolic product, probably in this way: (5.39).

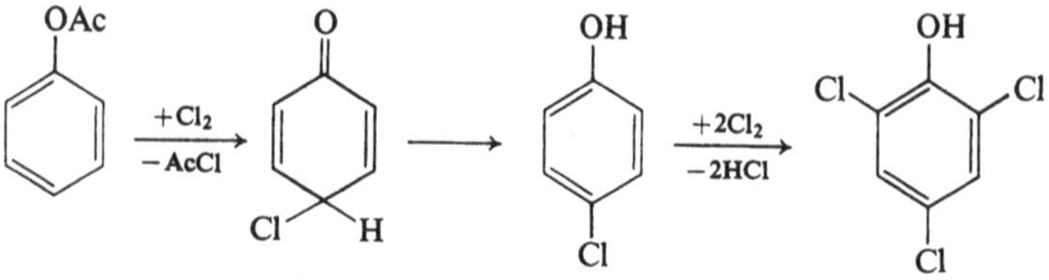

(5.39)

The tendency of phenols to give poly- rather than mono-substituted derivatives in halogenation results from another circumstance. As each

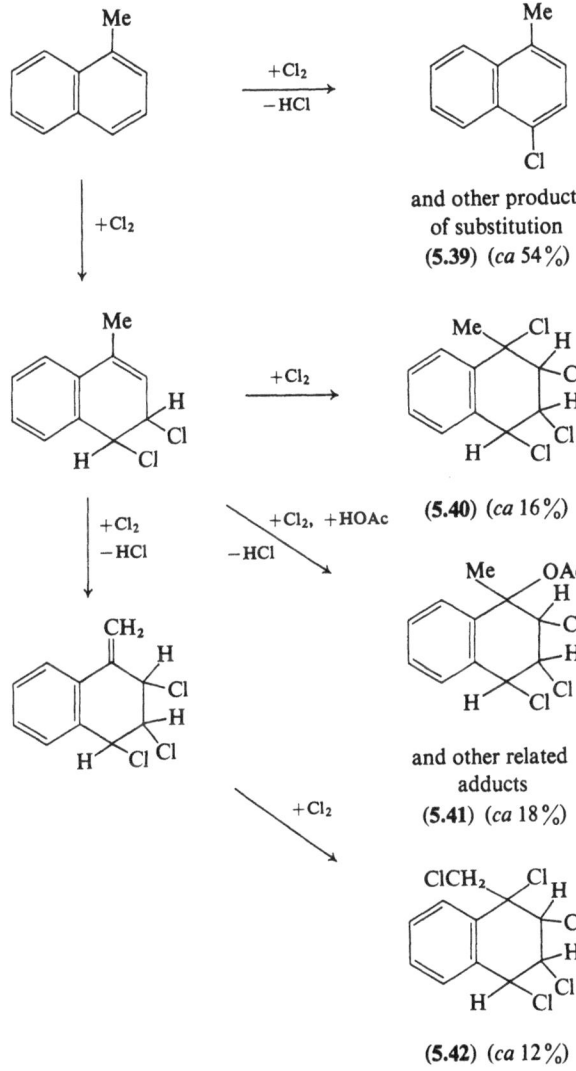

Fig. 5.6. Probable reaction paths to products in the reaction of 1-methylnaphthalene with molecular chlorine in acetic acid.

successive chlorine substituent is introduced into the benzene ring, the acidity of the phenolic hydroxyl group is increased. The tendency for the compound to react through the phenoxide ion, activated for electrophilic attack by the O⁻ substituent, is therefore enhanced.

It should be noted, however, that still more complex reaction paths

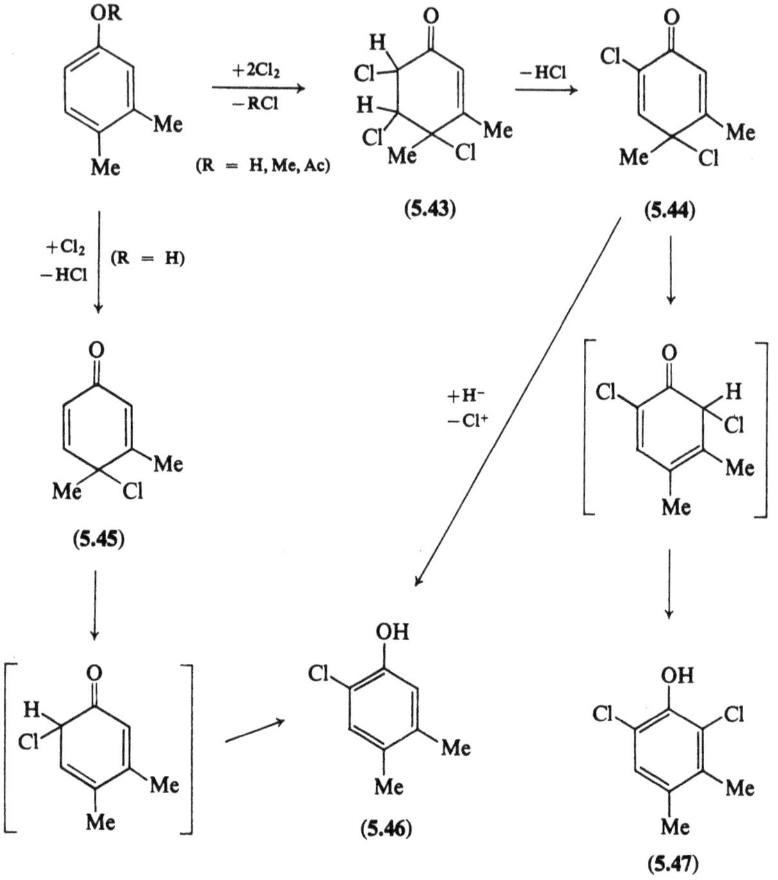

Fig. 5.7. Reaction pathways in the chlorination of 3,4-dimethylphenol and its *O*-derivatives in acetic acid. Square brackets indicate intermediates postulated but not isolated.

are concerned in some of these chlorinations, which have been reviewed in more detail elsewhere (de la Mare, 1971, 1974). Fig. 5.7 gives some of the reactions identified in the chlorination of 3,4-dimethylphenol and its *O*-derivatives; the pathways for these reactions may be contrasted with the apparently much simpler chlorinations of phenol and of anisole, from both of which can be obtained good yields of the simple products of substitution. Reactions readily recognisable in fig. 5.7 include the formation of products:

(*a*) of addition of chlorine, and replacement with rearrangement, not necessarily in that order, (**5.43**);

(b) of displacement with rearrangement, **(5.45)**;

(c) of (b), followed by double rearrangement, **(5.46)** formed via **(5.45)**;

(d) of (a), followed by elimination, **(5.44)**;

(e) of (d), followed by double rearrangement, **(5.47)**;

(f) of (d), followed by protodechlorination with rearrangement, **(5.46)** formed via **(5.44)**.

5.11 Product-forming sequences: addition–elimination pathways

Figure 5.7 illustrates, among other things, that addition–elimination sequences can be recognised as important in some chlorinations. The question of their general importance deserves some consideration, since the addition–elimination mechanism of substitution forms part of the history of chemistry, reviewed in the context of electrophilic substitution elsewhere (de la Mare, 1971). It was in vogue for many years; formed the issue of considerable controversy; and became partly neglected as the result of the recognition of the importance of carbocationic intermediates which in aromatic systems can lose a proton directly, with considerable gain in resonance energy. Adducts from chlorinations can, however, often be made to undergo eliminations; **(5.40)** illustrates that the

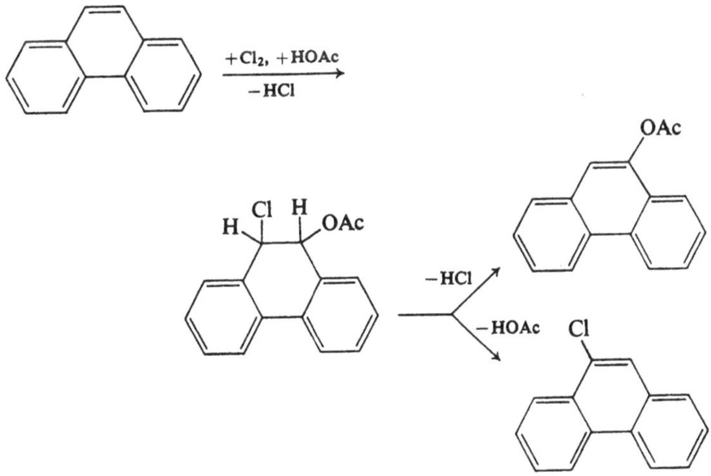

$$\text{(5.40)}$$

chloroacetoxy-adduct formed from phenanthrene and chlorine in acetic acid can undergo elimination in either of two directions, to give 9-chloro- or 9-acetoxy-phenanthrene. Conditions can be chosen for work-up of the reaction mixture to ensure the dominance of either of these modes of elimination, and it is obvious that the recognition of the formation of

9-acetoxyphenanthrene does not in this case indicate that an electrophilic acetoxylation has been under observation.

It often happens, when labile adducts are under consideration, that whether or not their participation can be detected depends on what techniques are available, or have been employed, for their detection. The formation or intervention of any particular adduct cannot be regarded as having been formally disproved until the properties of the hypothetical adduct are known or can be predicted with confidence.

5.12   Acetylenes and allenes

The considerations which affect the course of electrophilic additions to olefinic and to aromatic compounds apply also to the reactions of acetylenes and allenes. The chlorinations of these compounds, which are isomeric with conjugated dienes, have not yet been investigated in great mechanistic detail. For acetylenes, the usual second-order kinetic form $(-d[Cl_2]/dt = k_2[\text{Acetylenic compound}][Cl_2])$ seems to be dominant; the response in rate to structural change is in the direction expected for attack on the unsaturated compound by an electrophile; and the orientation of addition when solvent becomes incorporated is in the Markownikoff direction. In spite of the thermodynamic instability of acetylenes as manifested by their high heats of combustion, and of the accessibility of their unsaturation electrons, their reactions with chlorine are generally slower than those of the corresponding ethylenes; the transition states for the former compounds resemble vinyl cations, whereas those of the latter resemble carbocations, and the results show that the former are less easily attained from the initial state than are the latter. This may partly result from the fact that the nuclear charges on the carbon atoms in acetylenes, being very little shielded by bonding electrons, powerfully restrict the availability of the unsaturation electrons for co-ordination.

As with 1,3-dienes, the first stage of addition gives a product less reactive than the starting material, so the products of reaction with one molecular proportion of chlorine can easily be isolated. Substitution may accompany addition, as in the chlorination of phenylacetylene to give not only the two geometrically isomeric adducts, *cis-* and *trans-*1-phenyl-1,2-dichloroethylenes, but also 1-phenyl-2-chloroacetylene, (5.41), where the proton has been lost from the attacked carbon atom (Fahey, 1968).

$$\text{Ph.C}\vdots\text{CH} + Cl_2 \longrightarrow \text{Ph.C}\vdots\text{C.Cl} + HCl \qquad (5.41)$$

It has been assumed generally that these additions involve vinyl cations (e.g. structure (**5.48**)) as intermediates, or forms such as the ion-pairs or

$$R-\overset{+}{\underset{}{C}}=C\overset{H}{\underset{Cl}{\diagdown}} \qquad\qquad CH_2=\overset{+}{\underset{}{C}}-CH_2Cl \qquad\qquad \overset{Me}{\underset{Me}{\diagdown}}C=C=C\overset{H}{\underset{H}{\diagup}}$$

$$\textbf{(5.48)} \qquad\qquad\qquad \textbf{(5.49)} \qquad\qquad\qquad \textbf{(5.50)}$$

carbocationic complexes which might lead to them. Their subsequent reactions with nucleophiles do not seem to be stereospecific; but the extent to which they can exist in bridged forms is not known.

The double bonds in allenes (1,2-dienes) are quite rapidly attacked by electrophilic chlorine. Poutsma (1966, 1968; Poutsma and Kartch, 1966) has recorded the approximate scale of relative reactivities:

$$Me_2C:C:CH_2 \ (1000) > Me_2C:CH_2 \ (50) > C_2H_5CH:CH_2 \ (1) >$$
$$CH_2:CH.CH:CH_2 > C_2H_5C:CH.$$

Equations (5.42)–(5.44) show the products obtained from chlorine and some representative allenes under aprotic heterolytic conditions. The formation of propargyl chloride ($CH:C.CH_2Cl$) from allene, (5.42),

$$CH_2:C:CH_2 \quad\xrightarrow[\text{in } CH_2Cl_2 \text{ at } -30\,^{\circ}C]{+Cl_2}\quad \begin{array}{l} \nearrow\ CH_2:C(Cl).CH_2Cl \\[4pt] \searrow_{-HCl}\ \ CH:C.CH_2Cl \\[2pt] \text{and other products} \end{array} \qquad (5.42)$$

$$Me_2C:C:CH_2 \quad\xrightarrow[\ -HCl\ ]{+Cl_2}\quad \begin{array}{l} \nearrow\ Me_2C:C(Cl).CH_2Cl \\[4pt] \longrightarrow\ Me_2C(Cl).C(Cl):CH_2 \\[4pt] \searrow\ \overset{CH_2}{\underset{Me}{\diagdown}}C.C(Cl):CH_2 \end{array} \qquad (5.43)$$

$$Me_2C:C:CMe_2 \quad\xrightarrow[\ -HCl\ ]{+Cl_2}\quad \overset{CH_2}{\underset{Me}{\diagdown}}C.C(Cl):CMe_2 \qquad (5.44)$$

illustrates that terminal attack can be realised; here, it results in substitution with double-bond rearrangement. The simplest formulation of the carbocationic intermediate expected for this mode of attack is as a vinyl cation, (5.49), and the linear carbon skeleton of the original allene corresponds to the preferred linear geometry of such cations.

Central attack is also possible, as is shown by the formation of products of substitution with rearrangement from 1,1-dimethyl- and 1,1,3,3-tetramethylallene: (5.43) and (5.44). The product, (5.51), stoicheiometrically is a substituted allyl cation; but the geometry of the original allene, (5.50), together with the expected axial approach of the electro-

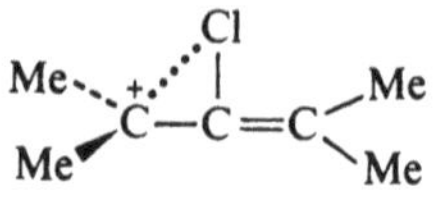

(5.51) 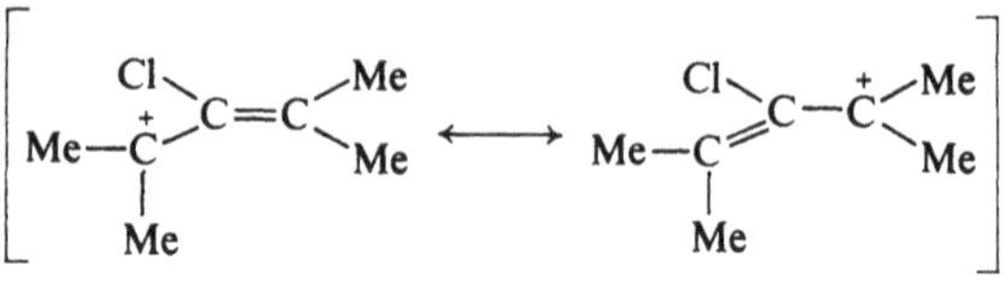   (5.52)

phile to the π-electrons of the double bond, ensures that it is formed in a geometry which does not correspond with that necessary for allylic delocalisation: (5.52). Stabilisation by this resonance, therefore, is only fully possible after the attack has been effected and geometrical reorganisation has ensued. Additional electron release (from the terminal methyl groups in the example cited) is necessary, therefore, to make electrophilic attack on the central carbon atom the dominant mode of reaction.

5.13 Summary

The general aims of this chapter have been:

(*a*) to illustrate the mechanistic generalisations mentioned in §5.1 for molecular chlorination, together with their wide applicability to simple olefinic compounds, to substituted olefinic compounds, to dienes and their analogues, and to aromatic compounds;

(*b*) to indicate the complex possibilities associated with the product-forming stages of the reaction path, thus illustrating for molecular chlorination most, if not all, of the ramifications available for carbocationic intermediates as set out in scheme 3.1.

It has been indicated that both kinetics and products give some indication of a special feature of molecular chlorination, namely that the chloride ion derived from the attacking electrophile may sometimes play a particular role, thus requiring its inclusion in the formulation of intermediates. For the latter, various geometrical forms have been considered (e.g. (5.17) and (5.21)); specific stereochemical evidence for the importance of the geometry of (5.21) has been mentioned. For simplicity, however, many intermediates have been shown in the derived form from which the chloride ion has been removed by heterolysis, and such intermediates have been represented sometimes in the fully bridged form, (5.15), sometimes in the partly bridged form (e.g. (5.16)), and sometimes in the 'open' form (e.g. (5.30)). Distinction amongst these becomes important only in some circumstances; and the evidence relating to these distinctions is most easily considered through the comparison of similarities and differences between the behaviour of different halogenating agents.

In chapters 6 and 7 chlorinations involving other sources of electrophilic chlorine, and the various brominating reagents including molecular bromine, will be discussed. In these subsequent chapters, olefinic, dienoid and aromatic compounds will all be considered; but no attempt will be made to illustrate all the accessible types of reaction path for every reagent. Instead, theoretically and practically important points of mechanistic difference will be stressed.

6 Chlorinations by chlorinating species other than molecular chlorine

'Distinguishing those that have feathers, and bite,
From those that have whiskers, and scratch.' (Lewis Carroll)

6.1 Introduction

This chapter is concerned with reagents in which the electrophilic chlorine is supplied from a bond other than the Cl–Cl bond. A wide variety of compounds can be used in this way; among them are included:

(*a*) the interhalogen compound, chlorine fluoride (Cl–F);

(*b*) compounds containing Cl–I bonds: benzene iododichloride (Cl–IPhCl), iodine trichloride (Cl–ICl$_2$), and related compounds;

(*c*) compounds containing Cl–O or Cl–S bonds, including hypochlorous acid (Cl–OH), chlorine monoxide (Cl–OCl), methyl hypochlorite (Cl–OMe), t-butyl hypochlorite (Cl–OCMe$_3$), chlorine acetate (acetyl hypochlorite, Cl–O.CO.CH$_3$), sulphuryl chloride (Cl–SO$_2$Cl), and their analogues;

(*d*) compounds containing Cl–N bonds, including chlorine azide (Cl–N$_3$), *N*-chloroacetamide (Cl–NH.Ac), phenyl dichloroamine (Cl–NPhCl), *N*-chloroacetanilide (Cl–N(Ac)Ph), *N*-chlorobenzene sulphonamide (Cl–NH(SO$_2$Ph)), and their analogues;

(*e*) the protonated forms of many of the above reagents: Cl–OH$_2^+$, Cl–NHR$_2^+$, and their analogues;

(*f*) compounds produced by reaction of a chlorinating species with a Lewis acid: Cl–SbCl$_4$, Cl–CuCl, and similar compounds.

Just as it was necessary to consider whether molecular chlorine reacts as an electrophile itself or by pre-equilibrium to some derived form, so with the reagents under discussion in this chapter it is necessary to establish whether or not they can act directly, and if so under what circumstances. Indeed, such considerations are particularly important because, if the reagent is provided in the presence of chloride ion (which often is difficult to exclude completely from the medium), the formation of molecular chlorine by such a reaction as that of (6.1) is often favoured thermodynamically, and reaction through molecular chlorine may prevail.

$$Cl—X + Cl^- \;\rightleftharpoons\; Cl—Cl + X^- \tag{6.1}$$

Only a few of the reagents to be considered have been investigated in sufficient detail to define adequately the mechanistic pathways available for their use. In the first part of this chapter, attention will be paid particularly to the kinetic or structural features which give information concerning the rate-determining stages of the reactions. Attention will be drawn, where this is appropriate, to features of the reactions which differentiate them from molecular chlorinations; some of these differences are helpful in providing evidence concerning the detailed mode of operation of both reagents compared.

6.2 Compounds having chlorine–halogen bonds

6.2.1 Chlorine fluoride. The primary mode of reaction of a diatomic interhalogen compound is determined by the relative electronegativities of the two halogens; so iodine chloride acts as an iodinating, and bromine chloride as a brominating species. Chlorine fluoride, then, provides electrophilic chlorine. Its behaviour has not been examined extensively; when it has been used, it has been prepared *in situ* by the reaction of chlorine with silver fluoride in an aprotic solvent in the presence of the olefinic compound to be chlorinated (Hall and Manville, 1969), and so it is not certain whether or not molecular chlorine was the actual electrophile. The product mixtures obtained from the reagent and D-glucal (see chapter 5, structure (**5.23**)) were somewhat different from those found with molecular chlorine, however, so it is possible that chlorine fluoride is the true reagent, giving predominantly but not exclusively *syn*-addition through intermediates analogous to those concerned in the corresponding additions initiated by chlorine and by fluorine (chapter 4).

6.2.2 Benzene iododichloride. The possible mode of action of this reagent has considerable theoretical and practical interest, because of the possibility that it could provide chlorine to a double bond through a cyclic transition state, (**6.1**), and thus provide an unusual example of synchronous addition. Although early studies suggested this as a real possibility, later investigations of kinetics and products have so far not produced any clear confirmatory evidence. Free-radical additions can certainly be realised (Lasne, Masson and Thuillier, 1972); under more

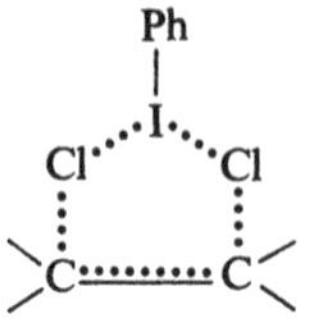

(6.1) 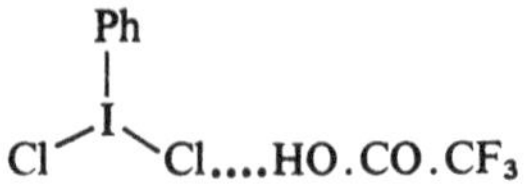   (6.2)

polar conditions, on the other hand, it often reacts by dissociation to chlorine and iodobenzene. In carbon tetrachloride containing trifluoroacetic acid, however, kinetic studies both of additions and of substitutions (Andrews and Keefer, 1960; Cotter, Andrews and Keefer, 1962) show that it can act itself as a source of electrophilic chlorine. The anion, which can be Cl^- or $CF_3.CO_2^-$ derived from the solvent, is then supplied separately to the carbocationic intermediate if required, and the sequence results, just as with molecular chlorine, in *anti*-addition to cyclohexene, and in a mixture of *syn-* and *anti*-addition to aryl-substituted olefinic compounds; and in substitution (6.2), with aromatic compounds. It

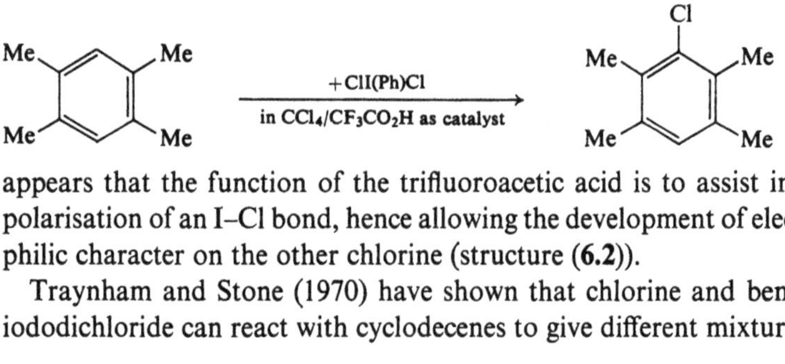

(6.2)

appears that the function of the trifluoroacetic acid is to assist in the polarisation of an I–Cl bond, hence allowing the development of electrophilic character on the other chlorine (structure (**6.2**)).

Traynham and Stone (1970) have shown that chlorine and benzene iododichloride can react with cyclodecenes to give different mixtures of

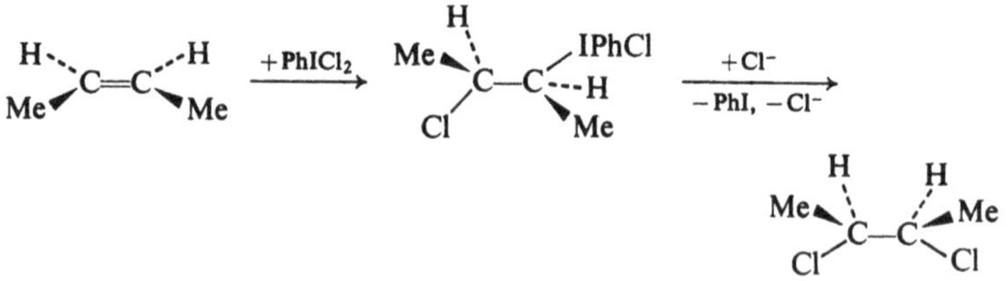

(6.3)

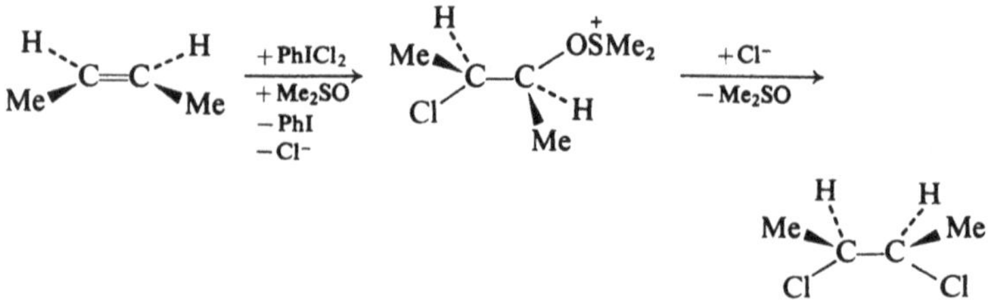

(6.4)

products of addition and rearrangement, a result which again suggests that the iododichloride can act as a chlorinating agent in its own right. Devillier and Bodot (1972) have recorded the interesting result that with simple alkenes, which normally add chlorine to give the product of *anti*-addition, benzene iododichloride in chloroform containing small amounts of dimethyl sulphoxide or of tetrahydrofuran gives some *syn*-addition. It was suggested that the reaction takes the course shown in (6.3); an alternative, which ascribes to the nucleophilic component of the solvent the specific role mentioned in chapter 5 (§5.6.5) is given in (6.4).

6.2.3   Iodine trichloride. Iodine is known to act as a catalyst for the chlorination of olefinic compounds, and is not consumed in the reaction. The kinetic form for the reaction of ethyl cinnamate with chlorine in carbon tetrachloride in the presence of a little iodine, (6.5), can be expressed as in (6.6). The sequence in which the reagents come together,

$$\mathrm{Ph.CH:CH.CO_2Et} \xrightarrow[\substack{\text{in CCl}_4;\, I_2 \\ \text{as catalyst}}]{+Cl_2} \mathrm{Ph.CH(Cl).CH(Cl).CO_2Et} \qquad (6.5)$$

$$-\mathrm{d[Olefin]}/\mathrm{d}t = k_2\,[\mathrm{Olefin}][\mathrm{Cl_2}][\mathrm{ICl}] \qquad (6.6)$$

and the precise way in which they function, is not known. Three possible representations of the rate-determining transition state are given in structures (**6.3**)–(**6.5**). In the first of these, chlorine and iodine chloride

(**6.3**)   (**6.4**)   (**6.5**)

are depicted as acting co-operatively, as electrophile and nucleophile respectively, from opposite faces of the ethylenic link. In the second, iodine chloride acts as a catalyst for removing chloride ion from a carbocationoid complex. In the third, preformed iodine trichloride acts as a chlorinating agent, the Cl–I bond becoming polarised in the direction necessary to form the iodine dichloride anion. These possibilities are not mutually exclusive, and all three modes, or any two of them, could operate together. Difficulties in distinguishing between isomeric transition states such as these, particularly when they are derived from three bulk components of the reaction medium, recur frequently in studies of halogenation, as will be seen in later sections and chapters.

Iodine trichloride has been used also as a bulk chlorinating species, and relatively high proportions of *ortho*-di-substituted derivatives have been reported (e.g. as shown in (6.7); Campaigne and Thompson, 1950). Such

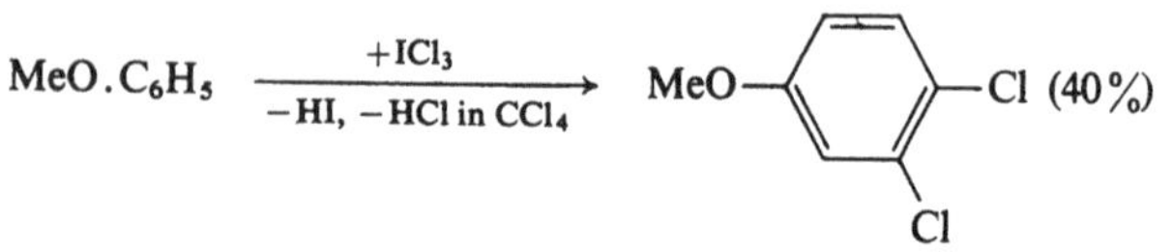

$$\text{MeO.C}_6\text{H}_5 \xrightarrow[\text{$-$HI, $-$HCl in CCl}_4]{\text{$+$ICl}_3} \text{MeO} \overset{}{-\!\!\!\bigcirc\!\!\!-} \text{Cl } (40\%) \qquad (6.7)$$

a result implies an unusual reaction path of some kind; but its exact nature is unknown, and an addition–elimination sequence may be involved.

6.3 Neutral species having chlorine–oxygen or chlorine–sulphur bonds

6.3.1 Introduction. The members of this group of compounds include a number of reagents which have been used preparatively, among them being hypochlorous acid, t-butyl hypochlorite and sulphuryl chloride. These compounds and their analogues are available equally for reactions with olefinic and with aromatic compounds; and such kinetic measurements as have been made reveal a wealth of possible reaction paths.

6.3.2 Hypochlorous acid. This reagent can be used for the preparation of chlorohydrins in aqueous solution: (6.8). Since it is a rather weak acid (pK_a, 7.47) it can be supplied in slightly alkaline solution, and is sometimes used in the mixture obtained by reaction of chlorine with alkalis, or through hydrolysis of such a salt as calcium hypochlorite which as usually obtained is a mixed salt of composition between $Ca(OCl)_2$ and $CaCl(OCl)$.

$$\text{Me.CH:CH}_2 + \text{ClOH} \longrightarrow \text{Me.CH(OH).CH}_2\text{Cl} \qquad (6.8)$$

Kinetic investigations, to which a number of groups of workers have contributed, have established that hypochlorous acid in very dilute aqueous solution will chlorinate olefinic compounds by a reaction having the simple kinetic form of (6.9). Presumably under these circumstances

$$-d[\text{Olefin}]/dt = k[\text{Olefin}][\text{ClOH}] \qquad (6.9)$$

the hypochlorous acid molecule itself attacks the double bond. At higher concentrations, however, a term involving the square of the concentration of hypochlorous acid becomes dominant, (6.10), and is presumed to

$$-d[\text{Olefin}]/dt = k'[\text{Olefin}][\text{ClOH}]^2 \qquad (6.10)$$

$$2\text{ClOH} \rightleftharpoons \text{ClOCl} + \text{H}_2\text{O} \qquad (6.11)$$

represent reaction through molecular chlorine monoxide, formed by (6.11). Likewise the formation of other new chlorinating species by the reaction of (6.12) has been recognised through the occurrence of kinetic terms as shown in (6.13).

$$ClOH + Nu^- \; \rightleftharpoons \; Cl\text{–}Nu + OH^- \qquad (6.12)$$

$$-d[\text{Olefin}]/dt = k''[\text{Olefin}][ClOH][Nu^-] \qquad (6.13)$$

Similar kinetic terms have been recognised for reaction of hypochlorous acid with aromatic compounds. It should be noted also that the kinetic possibilities have not been exhausted in the above description. Quite apart from those to be discussed later in relation to the formation and participation of positively charged analogues of these derivatives of hypochlorous acid, the actual rates of formation of the reacting electrophile from the bulk source of active chlorine in the medium can become rate determining, and then a new set of kinetic terms, independent of the concentration of organic substrate, contribute to the rate equation.

It is clear, therefore, that hypochlorous acid and many of its esters and ethers can act as sources of electrophilic chlorine. To determine the relative rates of attack by such reagents on the unsaturated compound, it is necessary to estimate the bulk concentrations of the various available reagents under conditions in which the second-order rate coefficients (k, k', k'' in (6.9), (6.10) and (6.13) respectively) can be measured. Only a few such comparisons are available; they suggest that the order of reactivity for reactions with olefinic compounds is:

$$Cl\text{–}OAc > Cl\text{–}OCl > Cl\text{–}Cl > Cl\text{–}OH$$

The same order probably holds for aromatic compounds, and is possibly to be associated with the relative strengths of the bonds to chlorine in these compounds (Evans, Lo and Chang, 1965), but these values are not known with great precision.

Not much is known to enable comparison of the orientational characteristics, or of the product-forming stages, in chlorinations of olefinic compounds initiated by these reagents; such information as exists suggests that in aqueous solution the intermediate stages are probably quite similar.

6.3.3 Chlorine acetate. Chlorination by chlorine acetate has some special features, which have been partly elucidated by examining its reactions with aromatic and with olefinic compounds in acetic acid and in aqueous acetic acid. The reagent can be prepared in various ways, including by the reaction of chlorine with mercuric acetate in acetic acid: (6.14).

$$Cl_2 + Hg(OAc)_2 \rightleftharpoons Hg(OAc)Cl + ClOAc \qquad (6.14)$$

$$ClOAc + H_2O \rightleftharpoons ClOH + HOAc \qquad (6.15)$$

Distillation under reduced pressure gives a solution of chlorine acetate. It can be recognised by its absorption spectrum; it is hydrolysed in water, (6.15), the position of equilibrium being so favourable for hydrolysis that in acetic acid its concentration is reduced to half by the presence of only about 0.1 per cent of water. Despite this, chlorine acetate is so much more effective than hypochlorous acid as an electrophile that it remains the active reagent even when the medium is made quite aqueous. It reacts as molecular chlorine acetate, with a response of rate to change in substituent rather less than that for molecular chlorine, but still quite considerable. The orientation for aromatic substitution and for addition to olefinic systems seems to be as expected for an electrophilic chlorinating agent.

For the chlorination of phenanthrene and of acenaphthylene, substitution and addition occur together; and from these results and from those obtained with olefinic substances it can be seen that the product-forming stages are quite different from those determined by molecular chlorine, even when these lead to the same products in the same solvent. Some of the important results are summarised in table 6.1. First, it can be seen that, if addition of chlorine to give dichlorides is compared with addition of chlorine acetate to give acetoxychlorides, the new reagent uniformly gives less *syn*-addition. Secondly, if the formations of acetoxychlorides from both reagents are compared, addition initiated by the new reagent uniformly gives more product of *syn*-addition. Thirdly, if additions to *cis*- and *trans*-olefinic isomers are compared, whereas chlorine gives widely different ratios of *erythro*- and *threo*-products, the new reagent gives a rather similar ratio, a result which suggests that each of the geometrical isomers may react through intermediates which have nearly the same geometry. Fourthly, whereas for addition initiated by chlorine added sodium acetate tends to increase the formation of *anti*-acetoxychlorides unless this value is already very large, no such tendency is obvious with the new reagent.

From these results it has been concluded that the greater reactivity of chlorine acetate than of molecular chlorine arises because chlorine acetate finds it easier to transfer positive chlorine completely to the olefinic compound, producing the carbocationic intermediate no longer encumbered by the presence of the anion. When an aryl-substituted

TABLE 6.1   *Products of addition of chlorine and of chlorine acetate to phenanthrene, to acenaphylene and to the methyl p-methylcinnamates*[a] *in acetic acid at 25 °C*

Compound	Added electrolyte ([NaOAc]/ mol dm^{-3})	Ratio, product of *anti*-addition to product of *syn*-addition		
		Dichloride adducts Reagent: Cl_2	Acetoxychloride adducts	
			Reagent: Cl_2	Reagent: ClOAc
Phenanthrene	—	0.27	2.4	1.0
Phenanthrene	NaOAc (0.1)	0.18	5.0	1.0
Acenaphthylene	—	0.71	3.8	3.0
Acenaphthylene	NaOAc (0.1)	0.73	7.0	2.0
Methyl p-methyl-cis-cinnamate	—	0.33	0.46	0.35
Methyl p-methyl-cis-cinnamate	NaOAc (0.1)	0.30	0.50	0.39
Methyl p-methyl-trans-cinnamate	—	0.53	Large	1.6
Methyl p-methyl-trans-cinnamate	NaOAc (0.1)	0.64	9.2	1.7

[a] The ratios of *erythro*-product: *threo*-product have the following values:
  cis-isomer, Cl_2, dichlorides, 0.33; acetoxychlorides, 0.46;
  trans-isomer, Cl_2, dichlorides, 1.89; acetoxychlorides, small;
  cis-isomer, ClOAc, acetoxychlorides, 0.35;
  trans-isomer, ClOAc, acetoxychlorides, 0.62.

olefinic compound is attacked, therefore, as in the examples given in table 6.1, bridging by the entering chlorine across the double bond is relatively ineffective; so completion of the reaction is fairly indiscriminate, giving *syn*- and *anti*-addition in a ratio which is affected by subtle environmental factors not yet fully understood, and in favourable cases giving approximately the same ratio of diastereoisomeric products from each of a pair of geometrically isomeric olefinic compounds.

  It is interesting that the synchronous pathway represented by the transition state shown in structure (**6.6**) does not seem to be important; it seems that chlorine acetate cannot easily complete the process of addition until much carbocationic charge has developed. Furthermore, the four-centred transition state shown in structure (**6.7**) is not available either, despite the occurrence of *cis*-addition in reactions of molecular chlorine, because the transition state again cannot be reached without the development of much carbocationic character. The special function of the chloride ion in determining the stereochemistry of addition is apparent in the comparison of chlorine and chlorine acetate as reagents.

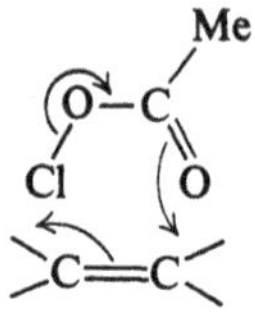

(6.6)

(6.7)

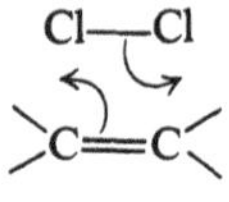

The details of the product-forming stages are sensitive also to structure of the substrate and to the solvent. For olefinic compounds in which bridging becomes important in the intermediate, the stereochemistry of addition of chlorine acetate becomes the same as that observed for molecular chlorine. Similarly, solvent effects on the products of addition of chlorine acetate to butadiene are in the expected direction; Heasley *et al.* (1972) have shown that the products are as shown in (6.16), and

$$CH_2{:}CH.CH{:}CH_2 \xrightarrow{+ClOAc} \begin{cases} Cl.CH_2.CH(OAc).CH{:}CH_2 \quad \text{(1,2-Adduct)} \\ Cl.CH_2.CH{:}CH.CH_2.OAc \quad \text{(1,4-Adduct)} \end{cases} \qquad (6.16)$$

that none of the product of 1,2-addition with reversed orientation could be detected. The 1,2:1,4-adduct ratio was 2.8 in n-pentane, 2.0 in acetic acid, and 1.3 in acetonitrile. The authors concluded that the carbocationic intermediate becomes more 'open', allowing more delocalisation of charge, the more ionising the solvent.

6.3.4 t-Butyl hypochlorite. t-Butyl hypochlorite is a readily available chlorinating agent, prepared by the reaction of chlorine with t-butanol, (6.17), in the presence of calcium carbonate to remove the hydrogen chloride as it is formed. It appears to be much more stable than most other

$$Me_3C.OH + Cl_2 \longrightarrow Cl.O.CMe_3 + HCl \qquad (6.17)$$

organic hypohalites are, and can be distilled and stored with reasonable safety. It has had some use for homolytic chlorinations, for which an initiator of free-radical chains is normally necessary. It has been found useful also for heterolytic chlorinations, which sometimes are carried out in aprotic solvents. Hydroxylic solvents such as acetic acid, however, give reactions which are more certainly heterolytic in character, and can involve substitution (6.18), addition (6.19), addition with incorporation of solvent (6.20), or addition with rearrangement (6.21).

$$Me.C_6H_5 + Cl.O.CMe_3 \longrightarrow Me.C_6H_4.Cl + Me_3C.OH \quad (6.18)$$

$$Ph.CH:CH_2 + Cl.O.CMe_3 \longrightarrow PhCH(O.CMe_3).CH_2Cl \quad (6.19)$$

$$CH_2:CMe.CH:CH_2 \xrightarrow[-Me_3C.OH]{+Cl.O.CMe_3, +HOAc} \begin{cases} Cl.CH_2.CMe:CH.CH_2.OAc \\ \\ Cl.CH_2.CMe(OAc).CH:CH_2 \end{cases} \quad (6.20)$$

$$p\text{-}R.C_6H_4.CMe(OH).CMe:CH_2 \xrightarrow[-Me_3C.OH]{+Cl.O.CMe_3} Me.CO.CMe(p\text{-}R.C_6H_4).CH_2Cl \quad (6.21)$$

These reactions all follow the pattern expected for heterolytic attack by electrophilic chlorine to give a chlorocarbocationic intermediate; but the nature of the products alone does not indicate whether or not the t-butyl hypochlorite molecule can attack the double bond directly, and many mechanistic discussions in the literature are manifestly incomplete. Kinetic evidence has, however, now been provided (M. A. Rosser, 1973: results personally communicated) to establish clearly that several of the reaction sequences hitherto known only as theoretical possibilities are available. The reaction of phenanthrene with t-butyl hypochlorite in acetic acid, for example, gives the 9-acetoxy-10-chloro-9,10-dihydro-phenanthrenes in the same ratio as for chlorine acetate under the same conditions; and the first-order kinetic form (rate independent of the concentration of phenanthrene, (6.22)) shows that the formation of chlorine acetate from t-butyl hypochlorite and acetic acid is rate deter-mining. The chlorination of toluene follows the same path, and proceeds at the same rate. The reaction with acenaphthylene under the same con-ditions, however, is much faster and is of the second kinetic order, (6.23), as has been established also (Kartashov, Pushkarev and Bodrikov, 1972) for the reaction of (6.21). The products from acenaphthylene are mainly acetoxychlorides, now formed in a ratio (3.8:1) significantly different from that (3.0:1) observed with chlorine acetate; and about 5 per cent of reaction giving t-butoxychloride adducts accompanies the main addition. It is apparent that this reaction involves rate-determining attack by some reagent other than chlorine acetate; almost certainly it is the t-butyl hypochlorite molecule, which clearly is much less effective as an electrophile than chlorine acetate or molecular chlorine.

$$-d[Cl.O.CMe_3]/dt = k[Cl.O.CMe_3] \qquad (6.22)$$

$$-d[Cl.O.CMe_3]/dt = k[Olefin][Cl.O.CMe_3] \qquad (6.23)$$

6.3.5 Sulphuryl chloride. Like t-butyl hypochlorite, sulphuryl chloride (SO_2Cl_2) is often thought of primarily as a homolytic chlorinating agent; but with aromatic compounds carrying electron-donating substituents, nuclear chlorination is generally smooth, (6.24), and the kinetics in

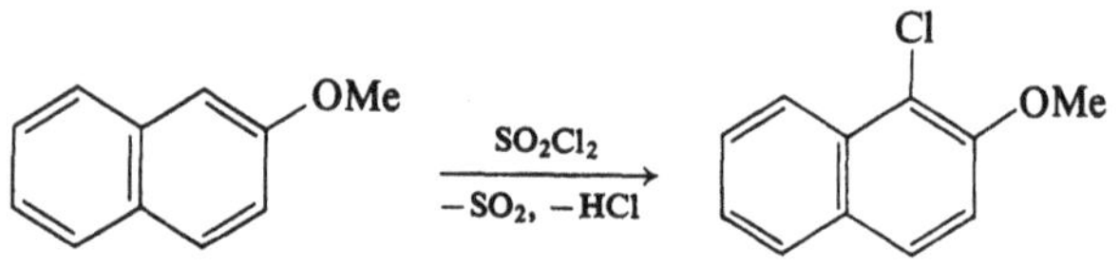

$$(6.24)$$

chlorobenzene and in other solvents indicate that the sulphuryl chloride molecule is the effective electrophile (Bolton and de la Mare, 1967; Bolton, 1968). The rate of reaction responds powerfully to electron release to the benzene ring ($\rho \approx -4$), and is increased by increase in the ionising power of the solvent, being some 1000 times greater in nitrobenzene than in chlorobenzene. The transition state therefore involves much carbocationic character and much development of charge from the formally neutral reactants.

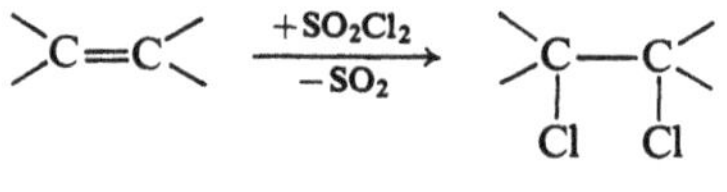

$$(6.25)$$

Addition, (6.25), though it has not been investigated kinetically, is also a known reaction. Thus from the reaction of naphthalene with sulphuryl chloride in liquid sulphur dioxide, 1-chloronaphthalene and a mixture of the naphthalene tetrachlorides typical of those formed in heterolytic addition of chlorine were identified. Still more complex product mixtures were formed from reaction with 1-methyl- and with 2-methyl-naphthalene (de la Mare and Suzuki, 1967). The details of these reaction paths are not known, though the literature contains many other examples of additions apparently initiated by molecular sulphuryl chloride.

Other sulphur-containing species, such as $ArSCl$, SCl_2 and $SOCl_2$, appear to act as donors of electrophilic sulphur rather than of electrophilic chlorine.

6.4   Protonated species having chlorine–oxygen bonds

6.4.1   Introduction. Protonation of any neutral species, ClOR, gives a positive species, $[ClORH]^+$: (6.26). Such species in principle could exist

$$ClOR + H^+ \rightleftharpoons [ClORH]^+ \qquad (6.26)$$

in isomeric forms (e.g. $[HCl.OR]^+$, $[Cl.O(H).R]^+$, $[Cl.O.RH]^+$), and further dissociation would give the chlorine cation, Cl^+ (resulting equilibrium, (6.27)). No values are known for the equilibrium constants of

$$ClOR + H^+ \rightleftharpoons Cl^+ + HOR \qquad (6.27)$$

such reactions in solution; small concentrations of the protonated forms and of Cl^+ are likely even under strongly acidic conditions, but the estimates mentioned in chapter 2 (§2.1) are highly uncertain because of unknown effects of solvent on the thermodynamic stability of Cl^+, which in the gas phase exists in a triplet state with two of its six valency electrons unpaired.

Evidence implicating such reagents as participants in chlorinations comes from studies of kinetics and of products. Acid catalysis indicates that such a reagent is under observation, provided that allowance is made for catalysis of any possible pre-equilibrium. It has been shown that reactions both of molecular chlorine acetate and of t-butyl hypochlorite are so catalysed, and it may be presumed that the ions $[Cl.O(H).CO.CH_3]^+$ and $[Cl.O(H).CMe_3]^+$ are involved.

6.4.2   Acid-catalysed chlorinations by hypochlorous acid; kinetic forms and structural effects. It has already been noted that molecular hypochlorous acid is a relatively ineffective chlorinating agent. In the presence of sufficient mineral acid, however, relatively unreactive aromatic compounds can be chlorinated. The reactions have been studied in water, and in various mixtures with organic solvents, often necessary to bring the organic substrate into solution. The kinetic form for these chlorinations is shown in (6.28); at acidities greater than about molar, the rate increases with acidity more rapidly than the stoicheiometric acidity, a result which is consistent with the view that the reaction involves the formation of a positively charged chlorinating species in pre-equilibrium.

$$-d[ClOH]/dt = k[ArH][ClOH]\cdot f[H^+] \qquad (6.28)$$

Acid catalysis aside, the most characteristic feature of these chlorinations is the small effective size of the entering chlorine. This can be illustrated by the partial rate factors shown in structure (**6.8**). For chlorination by molecular chlorine, the $\frac{1}{2}o\!:\!p$-ratio (0.75; structure (**6.9**)) falls below unity, probably because the chloride ion still associated in the

Me	Me	Me
134	617	306
4	5	
82	820	237
(**6.8**)	(**6.9**)	(**6.10**)
'Positive chlorine' in H$_2$O	Cl$_2$ in HOAc	ClOAc in aq. HOAc

transition state partly impedes *ortho*-substitution. For chlorination by 'positive chlorine', on the other hand, the $\frac{1}{2}o\!:\!p$-ratio (1.63) is greater than unity, a result which indicates that there is little if any steric hindrance impeding attack by the electrophile, and that the *ortho*-position is slightly activated preferentially by the inductive effect of the adjacent methyl group. These conclusions are made firmer by comparison with results for t-butylbenzene; here the steric effect of the t-butyl group is large for molecular chlorination, but is only just apparent for chlorination by positive chlorine.

Acidified hypochlorous acid, therefore, is delivering a reagent having minimal steric requirements, and this is true also for molecular chlorine acetate ($\frac{1}{2}o\!:\!p = 1.29$; structure (**6.10**)); it was concluded (p. 102–3) from other considerations that the latter reagent readily delivered positive chlorine to unsaturated systems. These results can be interpreted consistently as indicating that, in contrast with the significant role played by chloride ion from molecular chlorine, neither the water molecule derived from ClOH$_2^+$ nor the acetate ion derived from ClOAc suffer steric interference from the methyl group in the relevant transition states.

The reactions of a series of more reactive hydrocarbons have been examined kinetically in 96 per cent dioxan as solvent (de la Mare and Main, 1971; Burton *et al.*, 1972). The rates showed good correlation with the acidity of the medium as measured by the extent of protonation of *o*-nitroaniline. The products were mainly those of substitution, but the presence of small amounts of products of addition was detected for most of the substrates, which included the xylenes, mesitylene, triphenylene, naphthalene and phenanthrene.

For the most reactive compounds, reaction through molecular chlorine had to be avoided by the inclusion of silver perchlorate in the reaction mixture. It was found that the acid-catalysed reaction was further catalysed by silver chloride, and it was deduced that two modes of chlorination contributed, one involving $[ClOH_2]^+$ or its kinetic equivalent, and the other correspondingly involving $[ClAgCl]^+$.

The kinetic form of (6.26) ($f[H^+] = [H^+]$) had been recognised earlier as contributing to the chlorination of relatively reactive aromatic compounds (e.g. anisole) in dilute aqueous solution in the presence of silver perchlorate and acid. Considerable controversy developed from the recognition of a further kinetic term, (6.29), of zeroth order in aromatic compound. A number of possibilities have been suggested for the

$$-d[ClOH]/dt = k[ClOH][H^+] \qquad (6.29)$$

origin of this term, among them that it represents the rate of formation of Cl^+, of some more loosely solvated form of Cl^+ (e.g. $[Cl, 2H_2O]^+$), of $[HClOH]^+$, or of $[ClAgCl]^+$. It is at best a very minor component of a kinetically complex reaction.

Figure 6.1 illustrates through graph (*a*) that a good correlation is found between the rate of chlorination by 'positive chlorine' derived from acidified hypochlorous acid and theoretically derived (Streitwieser, 1961) parameters based on a transition state approximating in character to a simple chlorocarbonium ion (**6.11**). This result contrasts sharply with the

(6.11)

relatively poor correlation found when these 'positive chlorinations' are compared with the corresponding reactions of molecular chlorine, or when molecular chlorination itself is tested (graph (*b*)) against the same theoretical parameters. This result provides yet another indication that in molecular chlorination the chloride ion plays a special role in the transition state.

All the results so far noted in this section can be accommodated by assuming that the reactions pass through transition states having the composition [Unsaturated compound, $[ClOH_2]^+$] or [Unsaturated compound, $[ClAgCl]^+$], and it is usual to assume that the pre-equilibria of (6.30) or (6.31) precede the attack by electrophilic chlorine on the aromatic compound. Transition states of the same composition could,

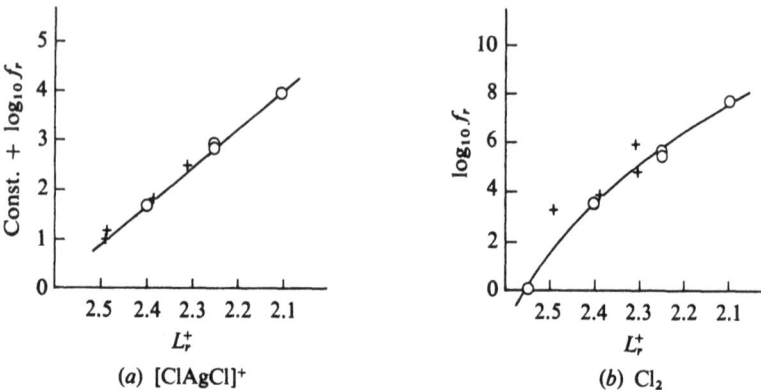

Fig. 6.1. Relationship between localisation energies (L_r^+) calculated for formation of intermediates of structures analogous to (**6.11**) and partial rate factors for reaction at 25 °C with 'positive chlorine', [ClAgCl]$^+$, in 96 per cent dioxan, graph (*a*), and with molecular chlorine in acetic acid, graph (*b*).

Points for reference compounds (*p*-xylene; the 2- and 4-positions of *m*-xylene; mesitylene; and, in (*b*), benzene also) are indicated by open circles and define the solid lines drawn; the curvature in (*b*) may be attributable to steric hindrance, which is known to be greater for molecular chlorination than for 'positive chlorination'. Points for the polycyclic hydrocarbons (the 1- and 2-positions of triphenylene; in (*a*) the 1- and 2-positions of naphthalene; and in (*b*) the 9-position of phenanthrene) are indicated by crosses; seriously aberrant points for molecular chlorination (graph (*b*)) are those for the 2-position in triphenylene and for the 9-position in phenanthrene. Data used are those given by Burton *et al.* (1972) and by Baciocchi and Illuminati (1962); the original papers should be consulted for further discussion.

$$\text{ClOH} + \text{H}^+ \ \rightleftharpoons \ [\text{ClOH}_2]^+ \qquad\qquad (6.30)$$

$$\text{ClOH} + \text{AgCl} + \text{H}^+ \ \rightleftharpoons \ [\text{ClAgCl}]^+ + \text{H}_2\text{O} \qquad\qquad (6.31)$$

however, be achieved equally by pre-association of the aromatic compound with hypochlorous acid and then a subsequent pre-equilibrium protonation of that complex or reaction of it with AgCl, followed by a rate-determining stage which would be indistinguishable in its results from that expected for the more usual description of the reaction path. No evidence exists to distinguish between these possibilities; the corresponding situation for bromination is discussed in chapter 7.

6.4.3 Orientation of addition to olefinic substances; the accompanying rearrangements, and substitutions accompanying addition, in the reactions of hypochlorous acid and acidified hypochlorous acid in aqueous solution. In this subsection attention is focussed particularly on the product-forming stages of these reactions. These in dilute aqueous solu-

TABLE 6.2 *Products of addition of hypochlorous acid to some derivatives* $(G.CH_2.CH:CH_2)$ *of propene*

		Reaction products; percentage of:		
		$G.CH_2.CH.CH_2$ $\quad\quad\;\;$│$\;$│ $\quad\quad$OH Cl	$G.CH_2.CH.CH_2$ $\quad\quad\;\;$│$\;$│ $\quad\quad$Cl OH	$HO.CH_2.CH.CH_2$ $\quad\quad\quad\;\;$│$\;$│ $\quad\quad\quad$G Cl
G	Solvent	(6.12)[a]	(6.13)[b]	(6.14)[c]
H	H_2O	90[d]	10	0
HO	H_2O	73[d]	27	0
Cl	H_2O	30[d]	66	4
Br	H_2O	32	40	28
I	H_2O	30	22	48
I	40% dioxan	31	39	30
I	70% dioxan	32	50	18

[a] Non-rearranged product of Markownikoff addition.
[b] Non-rearranged product of anti-Markownikoff addition.
[c] Rearranged product of Markownikoff-oriented attack.
[d] Addition of HCl to all these substrates gives 100 per cent of Markownikoff-oriented product.

tion seem to be reasonably independent of the source of electrophilic chlorine, so the intermediate stages will be represented as having the stoicheiometry [Unsaturated substance, Cl^+], the nucleophile originally associated with the electrophile (usually OH_2) being treated as unimportant.

Table 6.2 presents the most important results to which attention will be drawn; a more wide-ranging review has been given by Boguslavskaya (1972). Consideration of these shows first, that whereas addition of HCl gives no anti-Markownikoff product, addition of ClOH gives a significant proportion even for propylene; and with allyl chloride, anti-Markownikoff addition predominates but is not the exclusive mode of reaction.

Secondly, migration of the neighbouring group G, though not found for G = OH, is appreciable for G = Cl, and becomes progressively larger for G = Br and G = I, giving the product of structure (6.14). This mode of reaction does not, however, become the exclusive mode of reaction.

Thirdly, rearrangement competes more effectively with formation of other products, the more aqueous the solvent.

It can be presumed confidently that in an intermediate carbocation of composition $[Br.CH_2.CH^+.CH_2Cl]$ or $[I.CH_2.CH^+.CH_2Cl]$, the neighbouring bromine or iodine would take full control of the cationic centre by neighbouring-group interaction; as shown for example in structure (6.15), which would undergo ring opening to give no non-rearranged product of anti-Markownikoff addition. It can be presumed

$$CH_2—CH.CH_2Cl$$
$$\diagdown \diagup$$
$$I^+$$

(6.15)

also that a fully open carbocation of the same stoicheiometry would give on reaction with a nucleophile neither non-rearranged product of anti-Markownikoff addition nor rearranged product.

The reaction path adopted in these reactions is considered to be essentially as shown in fig. 6.2. The principal feature of the reaction path

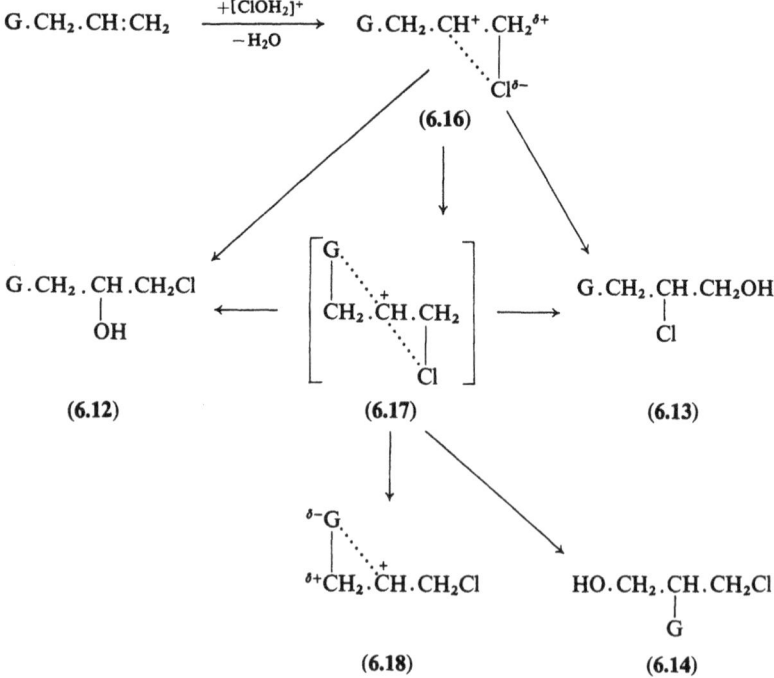

Fig. 6.2. Probable reaction paths in addition of hypochlorous acid to derivatives $(G.CH_2.CH:CH_2)$ of propene.

is that, in the first-formed intermediate (**6.16**), the bridging by entering chlorine must be significant to allow any formation of the product of anti-Markownikoff addition, but the intermediate cannot be symmetrically bridged, since otherwise much less product of Markownikoff addition would be expected. So it is represented as being bridged unsymmetrically by electrostatic interaction. Its reaction to give the Markownikoff product must be very rapid indeed; otherwise for G = Br or I, control of the carbocationic centre by the neighbouring group G would become exclusively predominant. In fact, rearrangement competes for these substituents; for G = Cl, a little migration of chlorine occurs, presumably through an intermediate or transition state having structure (**6.17**). The greater the solvation, and hence potentially the longer the life of the carbocationic intermediate, the greater the extent of rearrangement. The extent of rearrangement is not directly related to the expected driving power for neighbouring-group interaction, since the hydroxyl group does not migrate to the central carbon atom despite the fact that its power of interaction is greater than that of chlorine. This result supports the view that the interactions are not necessarily symmetrical.

The existence of neighbouring-group interaction in the carbocationic intermediates involved in these chlorinations receives support from study of substitution accompanying addition of hypochlorous acid. It has already been noted (§3.5.5; (3.18)) that electrophiles can react with iso-butene ($Me_2C:CH_2$) to give products of substitution with rearrangement; of the several reactions possible for isobutene and chlorine, the path shown in (3.18) predominates under aprotic conditions. In water, capture of the cationic intermediate by the solvent is naturally more prominent, but the ratio of direct substitution to substitution with rearrangement remains low. The results are summarised in fig. 6.3. The product of substitution with rearrangement (3-chloro-2-methylprop-1-ene (**6.21**)) is thermodynamically less stable than that of direct substitution (1-chloro-2-methylprop-1-ene (**6.20**)), and to account for the predominance of the former in the product, some special feature of the intermediate leading to it seems to be required; the open carbocation (**6.22**) would be expected to give predominantly the thermodynamically more stable olefinic product, by analogy with the Saytzeff orientation commonly observed in the unimolecular eliminations from alkyl halides. The most satisfactory interpretation of the formation of the product of substitution with rearrangement is that neighbouring-group interaction, (**6.23**), inhibits proton-loss from the 1-carbon atom, since the electronic movements necessary for this reaction (dashed arrow in (**6.23**)) are

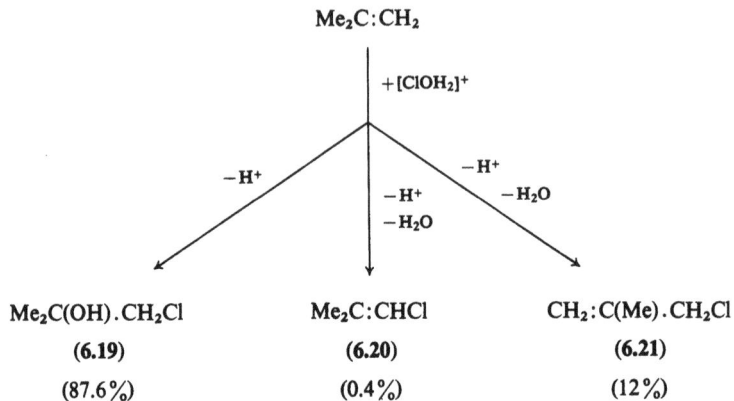

Fig. 6.3. Substitution accompanying addition in the reaction of hypochlorous acid with isobutene.

geometrically less satisfactory. The corresponding electronic movements leading to substitution with rearrangement (full arrow in **(6.23)**) are not subject to this stereo-electronic restriction. Reaction through the hypothetical symmetrically bridged cation **(6.24)** would even more strongly favour substitution with rearrangement; but then the isomeric chlorohydrin, $Me_2C(Cl).CH_2OH$, would be expected in the products, which do not contain it.

$$Me_2C^+.CH_2Cl$$

(6.22)   **(6.23)**   **(6.24)**

The products of chlorination of $[3-{}^{36}Cl]2,3$-dichloroprop-1-ene throw further light on the nature of the intermediates concerned in these reactions. The results are summarised in fig. 6.4. Two products of substitution are possible, differing only in the position of the isotopic label. One is that formed by direct substitution, **(6.27)**; the other, that of substitution with rearrangement, **(6.28)**. The latter predominates, a result which makes it abundantly clear that the proton-loss occurs before the intermediate **(6.25)** can achieve equivalence between the entering and the 3-chlorine atom, as it would have done in an 'open' carbocationic intermediate.

The reactions of hypochlorous acid and its esters with acetylenes and with allenes have been investigated only under preparative conditions,

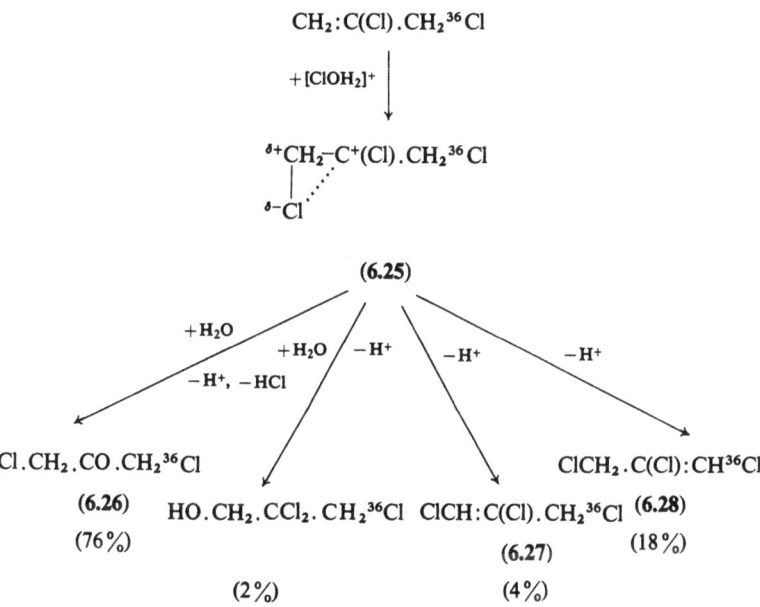

Fig. 6.4. Substitution accompanying addition in the reaction of hypochlorous acid with [3-^{36}Cl]2,3-dichloroprop-1-ene.

when the exact nature of the electrophile cannot be specified with certainty. The sequence shown in (6.32) (Stroh, 1962) gives evidence supporting the comment in chapter 5 (§5.12) that Markownikoff orientation is found with the former group of substances. The major pathway in the

$$R.C\vdots CH \xrightarrow{+2ClOH} R.C(OH)_2.CHCl_2 \xrightarrow{-H_2O} R.CO.CHCl_2 \quad (6.32)$$

reaction of hypochlorous acid with allene is as shown in (6.33), and illustrates circumstances under which attack at the terminal carbon atom to give a vinyl cation predominates for this compound; in chapter 5 it was noted that central attack can prevail with substituted allenes.

$$CH_2:C:CH_2 \xrightarrow{+ClOH} CH_2:C(Cl).CH_2OH \xrightarrow[-HCl]{+ClOH} ClCH_2.CO.CH_2OH \quad (6.33)$$

6.4.4 Acid-catalysed reactions of chlorine acetate and of t-butyl hypochlorite. Molecular chlorine acetate is so powerful an electrophile in its own right that identification of an acid-catalysed chlorination unambiguously involving a transition state having the composition [Unsaturated compound, ClOAc, H$^+$] has proved difficult. Additions of chlorine acetate to olefinic substrates and to aromatic compounds such as phenanthrene in acetic acid give product ratios which are very little affected by the presence of mineral acid.

Reactions having the required kinetic form are found for chlorinations by hypochlorous acid in aqueous acetic acid; these reactions may involve [ClOH$_2$]$^+$ or [ClOAcH]$^+$ (or their kinetic equivalents); when the solvent contains only a little water, it is probable that the latter is involved. Acid catalysis of the formation of chlorine acetate from hypochlorous acid and acetic acid has been noted (de la Mare *et al.*, 1960), but this of course does not establish whether or not a proton is concerned in the subsequent chlorination.

For t-butyl hypochlorite, a less effective molecular reagent than chlorine acetate, it should be easier to authenticate a chlorination involving the required transition state, now having the composition [Unsaturated compound, ClOCMe$_3$, H$^+$]. Acid-catalysed chlorinations by this reagent, however, are often performed in hydroxylic solvents. Under these circumstances, the formation of a new reagent (e.g. chlorine acetate; (6.34)) can become rate determining, and any observed acid catalysis may be of this reaction, rather than of the chlorination. Studies of the product ratios in additions and substitutions, however, (M. A. Rosser, 1973: results personally communicated) suggest that the reaction can occur not only in this way, but also through [Cl.O(CMe$_3$)H]$^+$ or its kinetic equivalent, since in some cases the product mixtures are different from those formed from chlorine acetate under the same conditions.

$$Cl.OCMe_3 + HOAc \quad \rightleftharpoons \quad ClOAc + HO.CMe_3 \qquad (6.34)$$

The chlorinations of acetylenes by alkyl hypochlorites have been studied also, again for preparative purposes; for example the formation of α,α-dichloroacetophenone from phenylacetylene and ethyl hypochlorite has been recorded (Stroh, 1962), and probably follows the sequence given in (6.35).

$$PhC\vdots CH \xrightarrow{+2EtOCl} PhC(OEt)_2.CHCl_2 \xrightarrow[-2EtOH]{+H_2O} Ph.CO.CHCl_2 \quad (6.35)$$

6.4.5   Chlorinations in sulphuric acid. Relatively unreactive aromatic substrates can sometimes be chlorinated conveniently by treating them with chlorine, sulphuric acid and silver sulphate. An organic solvent insoluble in sulphuric acid is often used to hold the chlorine, and the conditions are heterogeneous. It is clear that a powerful source of positive chlorine is being supplied, and (6.36) refers to a preparative

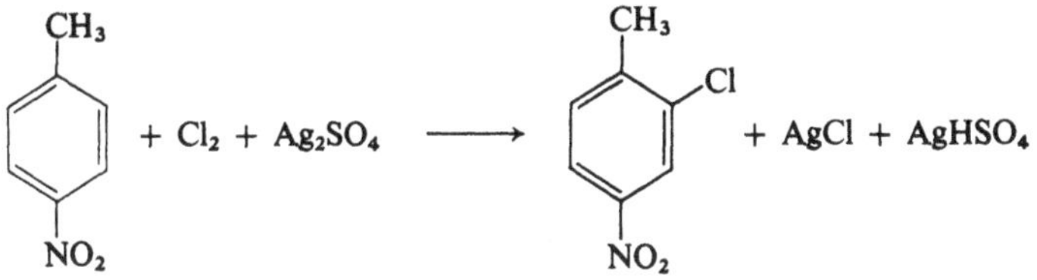

$$\text{(4-nitrotoluene)} + Cl_2 + Ag_2SO_4 \longrightarrow \text{(2-chloro-4-nitrotoluene)} + AgCl + AgHSO_4 \qquad (6.36)$$

example. It is not clear, however, what the effective reagent is; one possibility is $Cl.O.SO_2.OH$ or an isomeric or protonated form.

6.5   Species having chlorine–nitrogen bonds

N-Chloroamines of a wide variety of structural types are available, and have been used for substitutions and additions to unsaturated compounds. When these reagents are used in aprotic solvents, free-radical chain reactions can often be initiated.* Many of the known reactions, which have been reviewed comprehensively by Stroh (1962), could equally be heterolytic processes in which the N-chloro-compound acts as a source of electrophilic chlorine. Examples include substitution into acetanilide, (6.37), and addition to cyclohexene, (6.38). Mechanistic

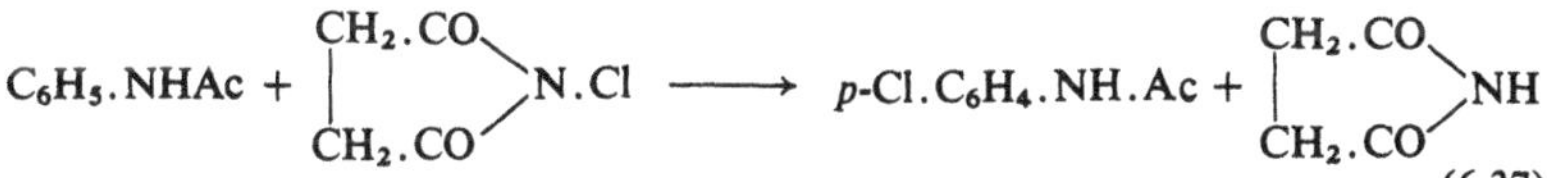

$$C_6H_5.NHAc + \underset{CH_2.CO}{\overset{CH_2.CO}{\diagdown}}N.Cl \longrightarrow p\text{-}Cl.C_6H_4.NH.Ac + \underset{CH_2.CO}{\overset{CH_2.CO}{\diagdown}}NH \qquad (6.37)$$

$$\text{(benzisothiazole)} \overset{CO}{\underset{SO_2}{\diagup}}N.Cl + \bigcirc \longrightarrow \text{(product)} \qquad (6.38)$$

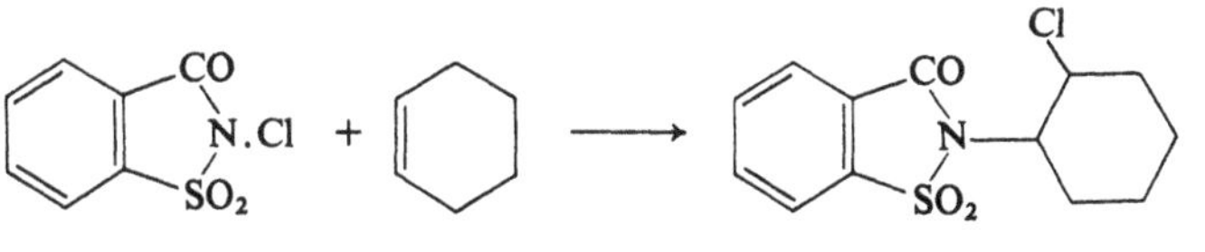

details have, however, not been established, and the similar reactions of nitrogen trichloride, which with hex-1-ene in methylene dichloride gives a mixture of dichloride and chloro-dichloramine adducts, may involve homolytic initiation (Kovacic, Lowery and Field, 1970).

Reactions which more probably are heterolytic in character, and take a somewhat unusual course, occur when aromatic compounds are treated with chloroamines (e.g. Me_2NCl, NCl_3) in the presence of an aluminium halide in nitromethane (Kovacic *et al.*, 1970). The effect of change in structure on the rate of reaction was found to be consistent with a process

---

* *Caution:* there are examples in the literature of dangerous explosions when such chlorinations have been attempted.

involving electrophilic attack on the aromatic compound. Anisole gives some chlorination, but for many aromatic compounds the main reaction is one of amination, which is believed to proceed by an addition–elimination sequence, and gives principally but not exclusively the *meta*-substituted derivative. A possible course for the reaction is given in (6.39). The effective electrophilic reagent may be an ion-pair, or a

$$\text{(6.39)}$$

(6.29)

complex $[Cl(N^+R_2)-Al^-Cl_3]$. Further reaction of this with the adduct **(6.29)** is probably inhibited by complexing between the latter and the aluminium halide.

As far as mechanistic investigations are concerned, the most extensively studied reactions of N-chloroamines are those in which they are used as a source of molecular chlorine, obtained by reaction, (6.40),

$$R_2N.Cl + HCl \rightleftharpoons R_2N.H + Cl_2 \qquad \text{(6.40)}$$

with hydrogen chloride. When the chlorine thus formed is used to effect a substitution or to add to an olefin with incorporation of a nucleophilic solvent, hydrogen chloride is regenerated, and so can be used in catalytic amounts to supply chlorine at low concentrations under mild conditions. When the amine formed (R_2NH, in (6.40)) has an appropriate structure, it can itself undergo chlorination. This happens in the well-known Orton rearrangement, of which an example is given in (6.41).

$$C_6H_5.N(Ac)Cl \xrightarrow{\text{HCl (as catalyst)}} o\text{- and } p\text{-Cl}.C_6H_4.NHAc \qquad \text{(6.41)}$$

One of the pieces of evidence supporting the above intermolecular route for the rearrangement is the fact that the $\frac{1}{2}o:p$-ratio for the rearrangement is the same as that for chlorination of acetanilide under the same conditions. There have been a number of attempts to establish the existence of other mechanisms, and one route which appears to give $\frac{1}{2}o:p$-ratios higher than those typically found for molecular chlorinations involves the treatment of aromatic amines with N-chlorosuccinimide in benzene (Neale, Shepers and Walsh, 1964). Haberfield and Paul (1965) have recorded the related rearrangement of N-chloro-N-methylaniline in carbon tetrachloride, again obtaining a high proportion of the *ortho-*

substituted product. The mechanism of the rearrangement occurring under these aprotic conditions is not known. One possibility is that the formation of the *o*-chloro-derivative involves in part an intramolecular rearrangement, (6.42), of the *N*-chloroamine or of its protonated form.

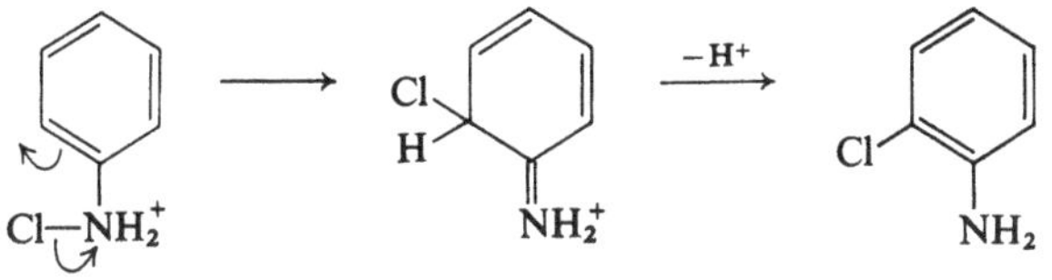

$$\text{(6.42)}$$

Another invokes the reverse direction of electron movement; Gassman and Campbell (1971) have investigated the rates of rearrangement of a series of *N*-chloroamines, $R.C_6H_4.N(Cl).CMe_3$, in ethanol buffered with acetic acid and sodium acetate as solvent, and find that the rate of reaction responds powerfully to electron release to the reaction centre ($\rho = -6.35$). This result suggests that the dominant path involves heterolysis to form a nitrenium cation: (6.43). Further work is needed to

$$R.C_6H_4.N(Cl).CMe_3 \xrightarrow{\ -Cl^- \ } R.C_6H_4.N^+.CMe_3 \qquad \text{(6.43)}$$

establish whether the mechanisms are the same in different solvents and under different conditions of acidity.

There seems little doubt that, when *N*-chloroamines are used in acetic acid with acids other than hydrogen chloride as catalysts, molecular species such as chlorine acetate may become involved ((6.44) for reaction

$$R_2N.Cl + HOAc \rightleftharpoons R_2NH + ClOAc \qquad \text{(6.44)}$$

in acetic acid). In a limited number of cases, it has been established that the protonated chloroamine itself can act as a chlorinating agent. Thus for the reaction of diethylchloroamine with phenol, the fact that the rate is independent of the acidity over a wide range of pH suggests that the reactants are the ions shown in (6.45). Similarly, Carr and England (1958)

$$Et_2NHCl^+ + OAr^- \longrightarrow Et_2NH + \text{Chlorinated phenols} \qquad \text{(6.45)}$$

showed that the *N*-chloromorpholinium cation (**6.30**) can act as a kinetic source of electrophilic chlorine for attack on the phenol molecule.

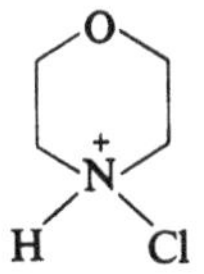

(6.30)

It can be presumed also, from the fact that amines can act as catalysts for the chlorination of relatively unreactive olefinic and aromatic substances, that equilibria of the type shown in (6.46) can produce new

$$C_5H_5N + Cl_2 \quad \rightleftharpoons \quad C_5H_5NCl^+ + Cl^- \qquad (6.46)$$

species capable of acting as sources of electrophilic chlorine. The preparative chlorination of acrylonitrile in pyridine, (6.47), may involve such a reagent.

$$CH_2:CH.CN + Cl_2 \quad \longrightarrow \quad Cl.CH_2.CH(Cl).CN \qquad (6.47)$$

Addition of chlorine azide to olefinic compounds seems to be analogous to additions of other compounds containing Cl–N bonds. Free-radical additions can be initiated, and may give the orientation of addition opposite to that obtained under heterolytic conditions. In the latter case, for example, the orientation of addition to styrene is as expected for a reaction initiated by electrophilic chlorine: (6.48) (Hassner and Boerwinkle, 1969).

$$Ph.CH:CH_2 + ClN_3 \quad \longrightarrow \quad Ph.CH(N_3).CH_2Cl \qquad (6.48)$$

6.6   Chlorination by metallic halides

6.6.1   Sequences initiated by electrophilic metallation. There are several ways in which metallic halides can participate in reactions leading to the chlorination of unsaturated compounds. In the first of these, the electrophilic metal atom attacks the double bond to give a complex, or an organo-metallic compound, which then can undergo further reaction with the resulting introduction of chlorine into the molecule. There are many reactions of this kind; (6.49) gives an example (Yakubovic and Motsarev, 1953). Limitations of space preclude further discussion.

$$Ph_2SiCl_2 \quad \xrightarrow[- PhSiCl_3]{+ SbCl_5} \quad PhSbCl_4 \quad \xrightarrow{- SbCl_3} \quad PhCl \qquad (6.49)$$

6.6.2   Heterolysis of chlorine–nucleophile bonds catalysed by Lewis acids. A second class of chlorinations catalysed by metallic halides involves the supply of chlorine in the presence of a metal halide, which assists the development of electrophilic character by aiding the heterolysis of the Cl–Cl bond (e.g. as in (6.50)). Kinetic investigation of these reactions is

$$Cl_2 + MX_n \quad \rightleftharpoons \quad [Cl^+MX_nCl^-] \qquad (6.50)$$

experimentally quite difficult, but the chlorinations of some aromatic hydrocarbons in nitrobenzene with aluminium trichloride have the kinetic form shown in (6.51), and show a substantial response of rate to electron release in the substrate (Caille and Corriu, 1969). There is not much information which fixes the exact nature of the reagent in such

$$-d[Cl_2]/dt = k[ArH] [Cl_2] \ (f[AlCl_3]) \tag{6.51}$$

chlorinations. Kovacic and Sparks (1960, 1961) have studied the products of chlorination of toluene and of the halogenobenzenes catalysed by a number of metallic halides, and have shown that the $\frac{1}{2}o$:p-ratios are on the whole lower than those found for the similar reactions involving only molecular chlorine, and considerably lower than those found for reagents like ClOH and ClOAc which act as efficient donors of positive chlorine. The results suggest that in the reactions catalysed by metallic halides, the effective chlorinating species are usually bulky, a result which is probably consistent with the formulation given in (6.50).

6.6.3   Chlorinations initiated by metal-containing electrophiles. Several types of metal halides can act as, or can supply, covalent chlorinating species still containing the metal. Antimony pentachloride may provide an example. Kovacic and Sparks (1960, 1961) have proposed that this reagent acts not through $[Cl^+SbCl_6^-]$ on the analogy of (6.50), but instead by providing electrophilic $[ClSbCl_3]^+$: (6.52). More fully investigated are

$$2SbCl_5 \ \rightleftharpoons \ [ClSbCl_3]^+ \ [SbCl_6]^- \tag{6.52}$$

the chlorinations catalysed by cupric chloride, interest in which arises because of its use in a technically valuable process whereby additions to olefinic compounds can be carried out without the prior formation of chlorine. In this process, the olefinic compound is passed over cupric chloride suspended on pumice at a temperature of about 320 °C, when a reaction such as that shown in (6.53) occurs. A mixture of air and hydro-

$$2CuCl_2 + CH_2{:}CH_2 \ \longrightarrow \ Cl.CH_2CH_2Cl + 2CuCl \tag{6.53}$$

gen chloride can then be used to regenerate cupric chloride for further use: (6.54). The reaction can also be carried out in methanol at 110–130 °C, when a mixture of the dichloride and the methoxychloride is

$$4CuCl + 4HCl + O_2 \ \longrightarrow \ 4CuCl_2 + 2H_2O \tag{6.54}$$

obtained, as in the corresponding reaction of molecular chlorine. The kinetics are complicated by equilibria involving cupric chloride, cuprous chloride and chloride ions; but they suggest that the reaction path is that shown in fig. 6.5 (Koyano and Watanabe, 1971). It is apparent that this

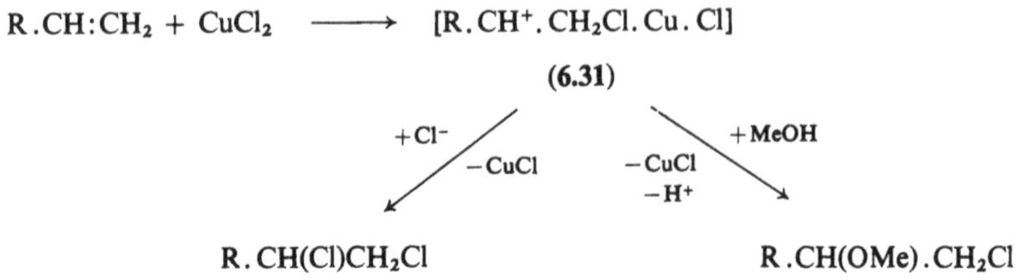

$$R.CH:CH_2 + CuCl_2 \longrightarrow [R.CH^+.CH_2Cl.Cu.Cl]$$

(6.31)

R. CH(Cl)CH$_2$Cl   R .CH(OMe).CH$_2$Cl

Fig. 6.5. Probable reaction path in chlorination by cupric chloride in methanol.

chlorination is considered to be essentially conventional in character, involving a new electrophilic chlorinating agent (Cl.CuCl) and a carbocationic intermediate (6.31).

Other higher metallic halides can probably react similarly, though when the details are fully worked out it may be useful to distinguish between those metals which have stable oxidation states one unit apart (especially certain transition metals), and those whose stable oxidation states are two units apart (as for example with the group V metals). Stannic chloride and molybdenum chloride have both been reported as capable of effecting chlorination. Akiyama, Horie and Matsuda (1973) have reported a novel result obtained by using antimony pentachloride in liquid sulphur dioxide. The products of reaction with cyclohexene under these conditions included 3-chlorocyclohexene (32 per cent), *trans*-1,2-dichlorocyclohexane (21 per cent), *trans*-1,3-dichlorocyclo-hexane (11 per cent), and *cis*-1,2-dichlorocyclohexane (3 per cent). All these products could be formed by way of the conventional type of carbo-cationic intermediate, though the 1,2-hydrogen shift needed to produce the 1,3-dichlorocyclohexane is unusual. But the reluctance of cyclohexene to give *cis*-addition in other chlorinations, including those with molecular chlorine and those catalysed by other metallic chlorides, makes it possible that *cis*-1,2-dichlorocyclohexane is formed in this reaction through a cyclic intermediate or transition state (e.g. (6.32)), as has been proposed also by Uemura, Sasaki and Okano (1971).

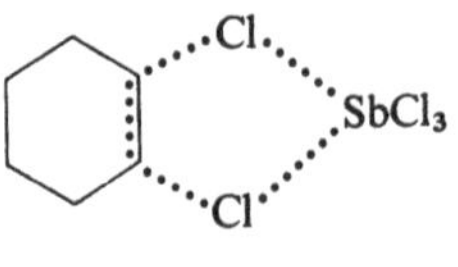

(6.32)

Another example of a chlorination in which the electrophilic reagent is a metallic halide involves the rearrangement of the carbocationic intermediate in the chlorination, (6.55), of 'hexamethyl-Dewar-benzene' (**6.33**) by auric chloride: (Hüttel, Tauchner and Forkl, 1972).

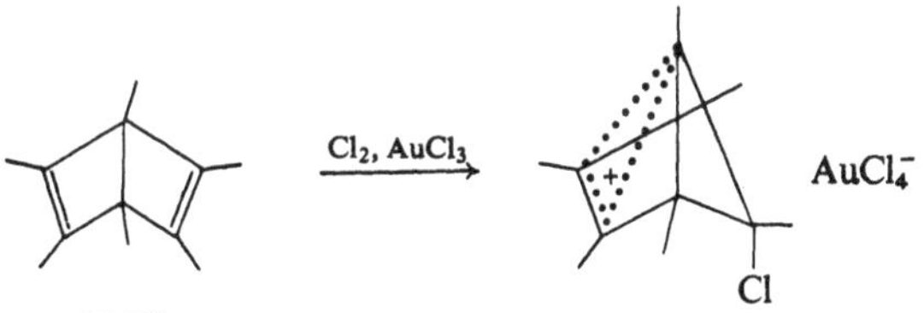

(**6.33**)

7 The bromination of unsaturated compounds

'The same law holds good for the Bromides.' (*O.E.D.* vol. I, p. 1125)

7.1 Introduction

The reactions of bromine and of other brominating agents with unsaturated compounds generally resemble those of chlorine; and the principles outlined in chapters 5 and 6 could equally well have been developed and exemplified by reference to the higher halogen. In this chapter, attention will be focussed on the major ways in which brominating species, including those of the type Br–X, differ in their behaviour from the corresponding chlorinating species.

These differences have several causes. Thus bromine is larger than chlorine, so steric effects on product formation are more evident; it bears a positive charge more readily, so reactions involving 'positive bromine' are more readily recognised; it forms weaker bonds to carbon, so molecular bromine is less reactive than molecular chlorine. When these bonds are formed, however, they undergo more rapid interchanges, effects of reversibility become more easily apparent, and the products are more frequently determined by thermodynamic control. Bromine also more strongly bridges between carbocationic centres, thus affecting the orientation, stereochemistry and potential for rearrangement of the intermediates. In general the kinetic behaviour, particularly for reactions of molecular bromine as compared with those of molecular chlorine, is more complicated, but the resulting product mixtures are simpler and cleaner.

The list of references includes a number of useful reviews; among them, those by Berliner (1964, 1966), Bradfield and Jones (1941), de la Mare and Bolton (1966), de la Mare and Swedlund (1973), Taylor (1972) and Williams (1941). These, and those cited in chapter 5 (§5.1) should be consulted for fuller documentation.

7.2 Reactions involving 'positive bromine'

Bromination of aromatic compounds by molecular hypobromous acid in water or in mixtures of water with organic solvents is slow, but the inclusion of mineral acid provides a very reactive brominating solution which, for example, quite rapidly attacks benzene and with sufficient acid will attack even so unreactive a compound as nitrobenzene. The kinetic form is that given in (7.1); the rate follows the concentration of hydrogen ion at low concentrations of acid; at higher concentrations it increases more rapidly than this, and in fact even more rapidly than the acidity as measured by the extent of protonation of a neutral base (Hammett's acidity function, h_0). Hexadeuterobenzene reacts at about the same rate as benzene, showing that the stage of proton loss in the substitution has not become rate determining. A reverse hydrogen–deuterium solvent isotope effect, $k_{D_2O}/k_{H_2O} = 2.2$, has been recorded.

$$-d[BrOH]/dt = k[ArH][BrOH](f[H^+]) \qquad (7.1)$$

The rate responds powerfully to electron-releasing substituents ($\rho^+ \approx -6.2$); and the effective reagent turns out to have relatively small steric requirements. Thus even t-butylbenzene gives about 38 per cent of the *ortho*-derivative; whereas in nitration it gives only about 11 per cent and in molecular bromination only 10 per cent. As a result, acidified hypobromous acid, or related sources of 'positive bromine', can be used much more effectively than molecular bromine, or the nitronium ion, for introducing a substituent adjacent to bulky groups in the aromatic ring, and for the preparative bromination of deactivated aromatic substrates.

Olefinic compounds such as allyl trimethylammonium perchlorate and the ethylene sulphonate ion (Kanyaev, 1961) have been shown also to react with hypobromous acid following the kinetic form given in (7.1); addition is the major reaction of the first of these compounds, and substitution to give $CH_2:C(Br).SO_3^-$ of the second.

The conventional interpretation of these results is that the reaction pathway starts with a pre-equilibrium protonation of hypobromous acid, followed by rate-determining attack by this or a derived positive species on the aromatic molecule. Gilow and Ridd (1973), however, have noted a difficulty, namely that the kinetic form has been shown to hold over too great a range of reactivity. If the positively charged reagent is the same over the whole range, then its concentration at the lowest acidities would be too low for the rate to be what is observed even if reaction occurred at every encounter of the hypothesised reacting species. It is suggested, therefore, that (at any rate for the more reactive aromatic compounds) the reagents must be brought together by complex-formation, giving the pathway shown in (7.2). It is still not certain, in the

$$\text{ArH} \underset{-\text{BrOH}}{\overset{+\text{BrOH}}{\rightleftharpoons}} [\text{ArH, BrOH}] \underset{-\text{H}^+}{\overset{+\text{H}^+}{\rightleftharpoons}} [\text{ArH, BrOH, H}^+] \xrightarrow[\text{determining}]{\text{Rate-}} \text{Products} \tag{7.2}$$

writer's opinion, whether this is the correct interpretation of the kinetic results; bromination through Br^+ or another positively charged species BrNuH^+ at the higher, and through BrOH_2^+ at the lower acidities would seem still to be possible. If it is correct, however, the results provide the first evidence that complexes of the type [ArH, BrNu] can play a role that is essential to the bromination. Even so, it is clear that the transition state for substitution is positively charged, has much carbocationic character, and is formed under circumstances in which substituents *ortho* to the site of substitution have a relatively small steric influence.

7.3 Kinetic forms, and structural effects, for reactions with molecular bromine

7.3.1 Reactions following second-order kinetics. The simple kinetic form of (7.3) has been established as being available for reaction of bromine with a wide variety of aromatic and olefinic substrates. To isolate this kinetic form free from complications, rather dilute solutions in highly ionising solvents are needed. The reaction rate is found to increase strongly with electron release from substituents suitably placed in the attacked molecule, a result which confirms that electrophilic attack to give an intermediate having carbocationic character is usually involved.

$$-\text{d}[\text{Br}_2]/\text{d}t = k_2[\text{Unsaturated compound}][\text{Br}_2] \tag{7.3}$$

Among the important series of aromatic compounds for which this kinetic term has been isolated is included the series of aromatic polycyclic compounds. In fig. 7.1 the rates of bromination as studied by Berliner and Altschuler (1966) are plotted against theoretical parameters which have been calculated on a model of the transition state having carbocationic character; they confirm that this is a good representation of the course of the reaction. Bromodesilylation, (7.4), and bromodeboronation, (7.5), have a similar kinetic form, though here the response of rate to change in substituent is rather less.

$$\text{Ar.SiMe}_3 + \text{Br}_2 \xrightarrow{+\text{H}_2\text{O}} \text{Ar.Br} + \text{Me}_3\text{SiOH} + \text{HBr} \tag{7.4}$$

$$\text{Ar.B(OH)}_2 + \text{Br}_2 \xrightarrow{+\text{H}_2\text{O}} \text{Ar.Br} + \text{B(OH)}_3 + \text{H}_2\text{O} \tag{7.5}$$

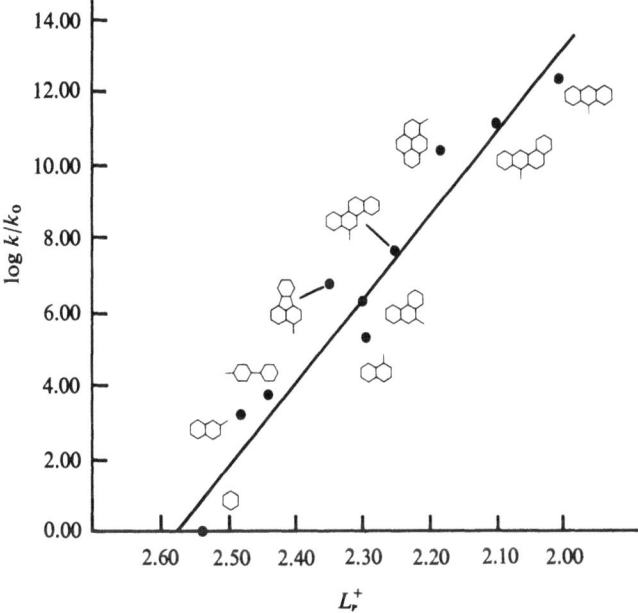

Fig. 7.1. Correlation of rates of bromination of polycyclic hydrocarbons with local-isation energies, L_r^+ calculated for formation of carbocationic intermediates by electrophilic attack on the positions indicated. (Reprinted with permission from Berliner and Altschuler (1966), *J. Amer. Chem. Soc.* **88**, 5837. Copyright the American Chemical Society.)

A reaction which follows the same kinetic form, but is subject to substituent effects of rather a different kind, is the bromination with rearrangement shown in (7.6) (Baciocchi and Illuminati, 1967). The

$$\text{(7.6)}$$

relative rates are given in table 7.1. The reaction is affected by the ionising power of the solvent in the expected way, being faster in the aqueous solvent. Bromine as a substituent shows its expected electron-withdrawing polar influence; presumably this is augmented slightly by steric hindrance, since the methyl and t-butyl groups slightly retard rather than facilitate the substitution. These rates can be regarded as reflecting the rates of *ipso*-attack with rearrangement.

TABLE 7.1 *Rates of bromination of 4-R-substituted 2,6-di-t-butylphenols in acetic acid and in aqueous acetic acid at 25 °C*

	$k_2/$l mol^{-1} s^{-1} in solvent:	
R	HOAc	98% HOAc
H	4.8	25
Me	1.3	7
t-Bu	1.3	6
Br	$<10^{-4}$	$<10^{-5}$

Kinetic and spectroscopic investigations of a number of other bromina-tions of phenols, naphthols and their derivatives have revealed the formation of dienones as intermediates through reactions following second-order kinetics. Thus bromodecarboxylations of substituted hydroxybenzoic acids, bromodesulphonations of some substituted phenol sulphonic acids, and the bromination of some naphthol sulphonic acids all take this course; the type of pathway is illustrated in (7.7).

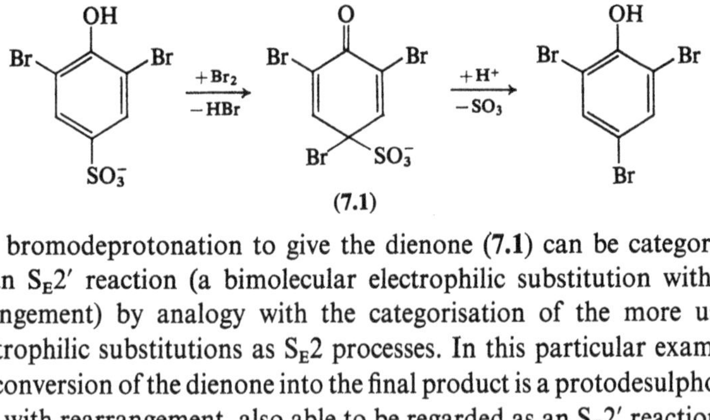

(7.7)

(7.1)

The bromodeprotonation to give the dienone (7.1) can be categorised as an S_E2' reaction (a bimolecular electrophilic substitution with re-arrangement) by analogy with the categorisation of the more usual electrophilic substitutions as S_E2 processes. In this particular example, the conversion of the dienone into the final product is a protodesulphona-tion with rearrangement, also able to be regarded as an S_E2' reaction.

Just as one or more intermediates may be involved in an S_E2 reaction, so also the S_E2' process may be more complex than is indicated in the sequence (7.7). A little attention has been given to the question of whether the proton loss from the phenolic hydroxyl group is part of the rate-determining stage of the reaction. For the bromination of phenol in acetic acid, the solvent deuterium isotope effect, k_2^{HOAc}/k_2^{DOAc}, has the value 1.9, a result which suggests that the breaking of the OH bond occurs during the formation of the transition state (de la Mare and Dusouqui, 1967).

The most extensive investigations of the effects of substituents on the rates of second-order additions of bromine available at present are those by Dubois and co-workers, who have used an electrometric method which allows the study of fast reactions in very dilute solution. Space allows discussion only of a limited selection from their findings. The rates of bromination of substituted styrenes (Dubois and Schwarcz, 1964), examined also by Pincock and Yates (1970), confirm that the reactions involve a carbocationic intermediate, with substantial response of the rate to change in the substituent (for $R.C_6H_4.CH:CH_2$, $\rho^+ = -4.5$ in MeOH, -4.3 in HOAc). Attempts have been made through studies of this kind to reach conclusions concerning the best model for the transition state. An important point at issue, for bromination as for chlorination (§5.3), is whether the bridged model (**7.2**) or a model based on attack proceeding simultaneously in separate processes at the two olefinic carbon atoms, (**7.3**) and (**7.4**), should be used.

| (7.2) | (7.3) | (7.4) |

For the related situation in substitution in benzenoid compounds, distinction between the two possibilities is easily made. *m*-Xylene, for example, is so much more reactive than *o*-xylene that it is obvious that substituents act additively in their effects on the activation of particular ring-positions and not on the activation of particular bonds, so the dissection of rates into partial rate factors for individual ring-positions is justified. The relevant comparison for olefinic compounds is to ask whether or not the β-methyl group is as effective in activating $Me_2C_\alpha H:C_\beta H.Me$ as the two α-methyl groups separately are in activating $Me_2C_\alpha:C_\beta H_2$, statistical factors being allowed for. If so, the groups act cumulatively in activating a bond (model, structure (**7.2**)); but if instead the two compounds have similar reactivity, then the effect of an activating methyl group is more powerful on positions β than on positions α to it, and a model based on structures (**7.3**) and (**7.4**) would be more appropriate.

Analysis of the results for any particular instance is not fully straightforward, for a number of reasons, one of which is that allowance for the difference of reactivities of *cis*- and *trans*-olefinic isomers is difficult. Dubois and Mouvier (1963, 1964, 1965) report the relative reactivities of methyl-substituted ethylenes given in table 7.2. The rate enhancement per methyl group is obviously not constant through the series; in the

TABLE 7.2 *Relative rates of bromination of ethylene and methyl-substituted ethylenes in methanol at 25 °C with added sodium bromide*

Compound	Relative rates of second-order bromination		
	Found	Calculated	
		Model 1[a]	Model 2[b]
(i) $CH_2:CH_2$	1	1	1
(ii) $Me.CH:CH_2$	6.1×10	4.1×10	6.0×10
(iii) *cis*-$Me.CH:CH.Me$	2.60×10^3	1.68×10^3	1.28×10^3
(iv) *trans*-$Me.CH:CH.Me$	1.68×10^3	1.68×10^3	1.28×10^3
(v) $Me_2C:CH_2$	5.41×10^3	1.68×10^3	5.8×10^3
(vi) $Me_2C:CH.Me$	9.1×10^4	6.9×10^4	7.6×10^4
(vii) $Me_2C:CMe_2$	9.2×10^5	28.3×10^5	16.5×10^5

[a] Calculated values for model 1, (7.2), were obtained by assuming that each methyl group activates the bond by a factor of 41.
[b] Calculated values for model 2, (7.3) and (7.4), were obtained by assuming that each olefinic carbon atom is activated, relative to a single position in ethylene, by a factor of 107 for a β-methyl group and 12 for an α-methyl group; so the calculated relative rates are for (ii), $0.5(107 + 12)$; for (iii) and (iv), 107×12; for (v), $0.5(107^2 + 12^2)$; for (vi), $0.5(107^2 \times 12 + 107 \times 12^2)$; for (vii), $107^2 \times 12^2$. Professor D. Hall is thanked for help with obtaining the best fit with this model.

second column are shown calculated results for the first model which give the best statistical fit on a free-energy basis; it is not as good as the fit that can be obtained by using the second model.

This conclusion agrees with that reached by Dubois and co-workers, and by Pincock and Yates (1970), for substituted styrenes. In this case, the large value of ρ^+ has been taken as diagnostic of reaction through a transition state which is unsymmetrical in character (model, structures

(7.5)

Steric repulsions between *ortho*-hydrogens prevent one or both phenyl groups from exerting their full normal electron-releasing power.

(7.3) and (7.4)). The type of test indicated above for the methyl group is not available for the phenyl group, since steric inhibition of resonance, (7.5), prevents use of the comparison of 1,1- with 1,2-, 1,1,2- and 1,1,2,2-aryl-substituted ethylenes (Ruasse and Dubois, 1972; Dubois *et al.*, 1972; Hegarty *et al.*, 1972).

7.3.2   Reactions following kinetics of order greater than one in bromine.
When the reactions of bromine with aromatic or with olefinic compounds in organic solvents are examined at concentrations of bromine in the range $0.001–0.05$ mol dm^{-3}, the incursion of a kinetic term, (7.8), of second order in bromine and third order overall becomes apparent.

$$-\text{d}[\text{Br}_2]/\text{d}t = k_3[\text{Unsaturated compound}]\,[\text{Br}_2]^2 \qquad (7.8)$$

This term becomes dominant in the higher part of the range, and extensive investigations of solvent and structural effects both for additions to olefins and for substitutions in aromatic compounds were made by Robertson and co-workers between 1939 and 1955. The work has been summarised (Robertson, 1955); the general characteristics of reactions following this kinetic form are very similar to those of second-order halogenations. The response of rate to electron release is high; a ρ^+-value of -12.1 has been given by Stock and Brown (1963) for aromatic substrates in aqueous acetic acid; and a similar conclusion reached by de la Mare (1949) for additions in acetic acid has been confirmed recently by Yates, McDonald and Shapiro (1973) by comparison of the ρ^+-values for second-order and third-order bromine additions to ring-substituted styrenes ($\rho^+ = -4.8$ and -4.6 respectively). Consequently the transition states must have considerable carbocationic character. This gives some help towards understanding the function of the extra molecule of bromine. It cannot, for example, be acting as a nucleophile in the final stage of an overall addition, for then the ρ^+-value would be expected to be less negative, and clear differences would be noted between addition and substitution. There are still, however, two different ways in which it could function. In the first, it could help to remove bromide ion from a complex between the unsaturated compound and bromine. In the second, it could form a dimer of bromine more polarisable than the monomer and hence more reactive. These alternatives are set out in (7.9) and (7.10) for

$$\text{ArH} + \text{Br}_2 \;\rightleftharpoons\; [\text{ArH}, \text{Br}_2]; \; [\text{ArH}, \text{Br}_2] + \text{Br}_2 \longrightarrow \text{Products} \qquad (7.9)$$

$$\text{Br}_2 + \text{Br}_2 \;\rightleftharpoons\; \text{Br}_4; \; \text{Br}_4 + \text{ArH} \longrightarrow \text{Products} \qquad (7.10)$$

aromatic substitution. It will be clear from study of chapter 2 that either of these possibilities makes chemical sense, and in fact no distinction can at present be made between them. It is possible, in fact, that both paths are acting together, and the likelihood of this is perhaps enhanced by the fact that kinetic terms of still higher order in bromine have been detected for the reactions of some compounds at still higher concentrations. The effectiveness of catalysts such as iodine bromide and zinc chloride in promoting electrophilic brominations and reducing the kinetic order with respect to bromine, reviewed by Taylor (1972), can probably be accommodated with either type of sequence, but perhaps more satisfactorily by the first. (See note added in proof, p. 158.)

7.3.3 Bromide-catalysed brominations. Associated with the kinetic terms in the kinetic equation for bromination discussed in the two foregoing subsections, a third has been recognised by a number of groups of investigators. It takes the form of (7.11) (which can be written also as in (7.12), where K_f is the formation constant of Br_3^- from Br_2 and Br^-) and is especially important for olefinic and acetylenic compounds.

$$-d[Br_2]/dt = k_{Br^-} \text{[Unsaturated compound]}[Br_2][Br^-] \qquad (7.11)$$

$$-d[Br_2]/dt = k_{Br^-} \, K_f^{-1} \text{[Unsaturated compound]} \, [Br_3^-] \qquad (7.12)$$

There are a number of ways in which such a kinetic term can arise. In the first, the bromide ion provides the nucleophilic partner in the addition as part of the rate-determining stage; either synchronously, (**7.6**), or after pre-association of the unsaturated compound and bromine. It would be characteristic of reaction by either of these routes that increase

(7.6)

in the concentration of bromide ion in the reaction medium would increase the proportion of *trans*-dibromides at the expense of other products, including those of incorporation of solvent. It would be expected also that structural effects on reaction rate would be smaller, in the sense that the response in rate to electron release would be less than in uncatalysed molecular bromination. It might still, however, be in the direction expected for electrophilic attack on the unsaturated compound, provided that bond formation to the bromine molecule had developed more completely than bond formation to the bromide ion in the transition state.

The kinetic and structural features of reaction by this mechanism were exemplified clearly by P. W. Robertson and his co-workers (reviewed by de la Mare, 1949; de la Mare and Bolton, 1966) for reactions of such compounds as $CH_2\!:\!CH.CH_2Cl$ and $CH_2\!:\!C(CH_2Cl)_2$ in acetic acid as solvent. More recent studies of kinetics and products from reactions in this and in other solvents have led to the general acceptance of this mode of function of bromide ions (Fahey, 1968; Rolston and Yates, 1969; Dubois and Huynh, 1971), though distinction between synchronous and stepwise pathways is a matter of unresolved controversy.

The reactions of bromine with alkyl- and aryl-substituted acetylenes have been investigated carefully by Pincock and Yates (1970). In acetic acid, these reactions show the general characteristics noted for the corresponding more rapid additions to olefinic substances. The three kinetic terms, now shown associated in (7.13), can all be recognised, and

$$-d[Br_2]/dt = (k_2[Br_2] + k_3[Br_2]^2 + k_{Br^-}[Br^-][Br_2])[\text{Acetylene}] \qquad (7.13)$$

their separate characteristics can be identified. Values of k_2 for ring-substituted phenylacetylenes are correlated well with values of σ^+, and the high ρ^+-value (-5.2) suggests that a carbocationic intermediate (probably a vinyl cation) is involved. Both *cis-* and *trans-*olefinic dibromides are obtained; the bromo-acetates which accompany them, (7.14), are entirely of the Markownikoff type. Product composition

$$\text{Ph}.\text{C}\!:\!\text{CH} \xrightarrow[-\text{HBr}]{+\text{Br}_2, +\text{HOAc}} \text{Ph}.\text{C(OAc)}\!:\!\text{CHBr} \qquad (7.14)$$

responds to the addition of electrolytes to the medium in a complicated way which suggests that the product-forming sequence involves ion-pair intermediates.

The bromide-catalysed mechanism gives only the *trans-*olefinic dibromide, and the response of rate to change in substituent is less pronounced $(\rho^+ = -1.9)$, and is not linear with σ^+. The latter result could be interpreted in terms of a changing transition-state structure; perhaps with simultaneous involvement of synchronous and stepwise processes.

7.3.4 Bromination by the tribromide ion.

7.3.4 Bromination by the tribromide ion. The kinetic term of (7.11) or its equivalent, (7.12), can arise in another way, namely through the tribromide ion attacking the unsaturated molecule. Two ways in which it could do this have been recognised. For unsaturated compounds deactivated by the presence of electron-withdrawing groups, Br_3^- could act as a nucleophile, as was proposed by McDonald, Milburn and Robertson (1950) for addition to methyl vinyl sulphone. This compound reacts with bromine less rapidly than does vinyl bromide in the absence of added bromide ions, but more rapidly in their presence, a result which suggests the incursion of a new mechanism (see also de la Mare and Bolton, 1966). Probably these reverse structural effects could equally well be accommodated by a mechanism involving a transition state like that shown in structure (**7.6**), but dominated by the nucleophilic rather than by the electrophilic component of the addition in either a synchronous or a stepwise process.

For activated unsaturated compounds, on the other hand, the tribromide ion could be envisaged as acting as a rather ineffective electrophile. Evidence for this type of behaviour has been adduced for bromination of olefinic compounds by Dubois and Huynh (1971), and for aromatic compounds by Berliner and Beckett (1957). Further details of these pathways have not been elucidated, but it is clear that attempts to interpret the various bromide-catalysed mechanisms are complicated by the availability of multiple pathways even for the operation of this single kinetic term.

7.3.5 Acid-catalysed brominations. Additions of bromine to $\alpha\beta$-unsaturated carbonyl compounds are subject to catalysis by acid, and particularly by hydrogen bromide. The reactions tend to be autocatalytic, and it has been speculated in the past that protonation of the carbonyl group might be followed by nucleophilic attack by Br_2 or by Br_3^- on the double bond, the final result being *anti*-addition of bromine. For the bromination of mesityl oxide in acetic acid, it is likely (Robertson and Swedlund, 1974) that a better interpretation involves electrophilic addition, (7.15), to an activated alkene (e.g. structure (**7.7**)) formed by

$$Me_2C:CH.C(Me):O \xrightarrow{\ +HBr\ }$$

$$Me_2C:CH.C(Me)(OH)Br \xrightarrow[-HBr]{\ +Br_2\ } Me_2C(Br).CH(Br).C(Me):O \quad (7.15)$$

$$(\textbf{7.7})$$

1,2-addition across the $C{=}O$ bond. Further work on these systems is desirable; it is also not yet clear to what extent homolytic processes can be concerned in these sequences. Related problems arise in relation to the bromination of quinones (Rothbaum, Ting and Robertson, 1948).

7.3.6   Primary kinetic isotope effects, and reversibility, in bromine substitutions. Although many bromine substitutions resemble chlorinations in showing no significant primary kinetic isotope effects, so that proton loss has made little progress in the transition state, two different types of structural situation can alter this position. The first provides steric overcrowding at the reaction site, as in the molecular bromination of 5-t-butylhemimellitene (**7.8**), for which a primary kinetic isotope effect, $k_H/k_D = 2.7$, has been recorded in nitromethane as solvent (Baciocchi *et al.*, 1967).

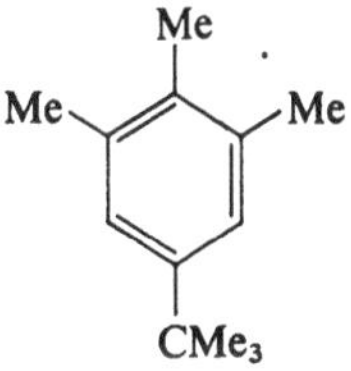

(7.8) 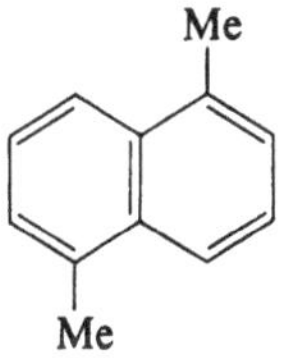   (7.9)

Berliner *et al.* (Berliner, Kim and Link, 1968; Kim *et al.*, 1970) have illustrated how important this fact can become in altering the orientation of substitution, from study of the rates of bromination of the methyl-substituted naphthalenes. The proportion of substitution in the 2- and 3-positions of 1,5-dimethylnaphthalene (**7.9**) is greatly in excess of what would be expected. It was shown that this results in part from the incursion of addition–elimination pathways, and in part from a change towards the mechanism in which the stage of proton-loss has become partly rate determining. Attack on the 4-position is sterically affected by the adjacent methyl group, so that proton loss becomes rate determining for reaction at this position only: the proportion of reaction at other positions is consequently unexpectedly high.

The second structural feature favouring observation of a significant primary kinetic isotope effect is electron release to the reacting centre. Farrell and Mason (1963) found that the value of k_H/k_D for the *ortho*-bromination of *N,N*-dimethylaniline was 2.6; the corresponding ratio for *para*-bromination was 1.0, and the detailed mechanism of substitution at the two positions is obviously different.

The observation of a significant primary kinetic isotope effect for any particular halogenation does not in itself allow distinction to be made between a synchronous mechanism, (7.16), and the two-step process involving a carbocationic intermediate, (7.17) and (7.18). If, however,

$$ArH + X_2 \;\rightleftharpoons\; [TS_A] \;\longrightarrow\; ArX + X^- + H^+ \qquad (7.16)$$

$$ArH + X_2 \;\rightleftharpoons\; [ArXH]^+ + X^- \qquad (7.17)$$

$$[ArXH]^+ \;\rightleftharpoons\; [TS_B] \;\longrightarrow\; ArX + H^+ \qquad (7.18)$$

k_H/k_D can be shown to vary with the concentration of halide ions, being greater at higher concentrations of X^-, the two-stage process can be regarded as established. For, in the synchronous mechanism, no such change would be predicted; but in the second, an increase in the concentration of X^- will increase the rate of return of the carbocationic intermediate $[ArXH]^+$ to starting material, and so may allow the stage of proton loss to make a greater contribution to the reaction rate. This criterion has been applied by Baliga and Bourns (1966) to the bromodeprotonation of sodium *p*-methoxybenzene sulphonate (**7.10**) and, by the use of sulphur kinetic isotope effects, to its bromodesulphonation also. Christen *et al.* (1962; cf. Christen and Zollinger, 1962) have used similar criteria in relation to the bromination of the disodium salt of 2-naphthol-6,8-disulphonic acid (§7.3.1). (See note, p. 158.)

(7.10)   (7.11)

Orientational aspects of assistance to departure of the leaving group have been indicated through studies by Reich and Cram (1969) of the bromination of [2,2]-paracyclophanes. Here, after electrophilic attack upon one ring, it is believed that the proton is transferred internally to the other, from which a proton is finally lost. In the example shown, (**7.11**), the carbomethoxy-group assists this transfer, and so promotes substitution in the 'pseudo-*gem*-' position shown by the arrow.

Under certain conditions it has been shown also that brominations can become reversible. Thus O'Bara, Balsley and Starer (1970) have found that hydrogen bromide catalyses the isomerisation and disproportionation of *p*-bromophenols in chloroform at 25 °C; 4-bromo-3,5-xylenol, for example, rapidly gives a mixture containing an excess of 2-bromo-3,5-xylenol, together with small amounts of debrominated and dibrominated products. Figure 7.2 shows the reaction path proposed for the slower reaction of *p*-bromophenol; electrophilic proton donation is

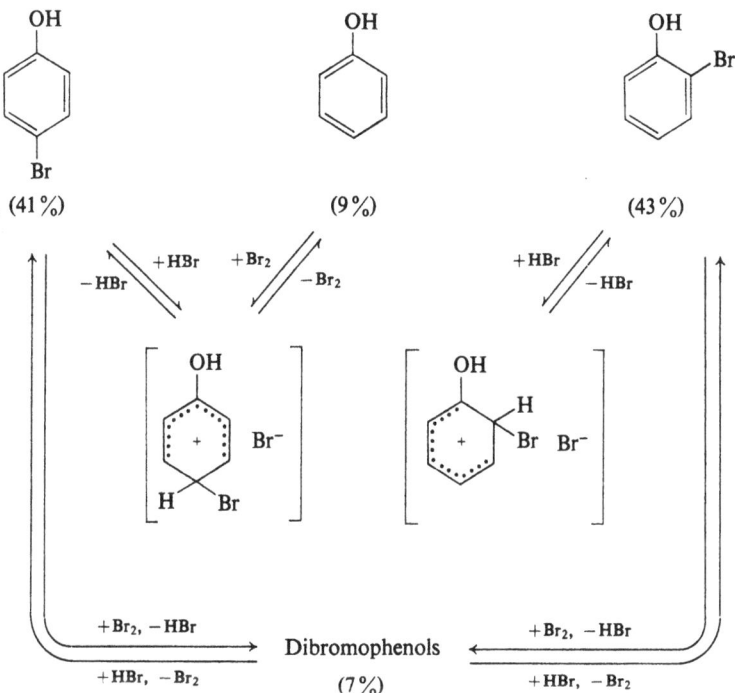

Fig. 7.2. Debromination, rebromination and disproportionation in the reaction of p-bromophenol with hydrogen bromide in chloroform at 25 °C. Percentages given in brackets are those observed after the reaction of p-bromophenol (1.16 M) in chloroform saturated with HBr at 25 °C after 40 days.

shown by the effect of substituents on the rate of isomerisation and is assisted by nucleophilic attack by bromide ion on the bromine substituent. Hydrogen chloride was shown to be a much worse catalyst, and toluene p-sulphonic acid was ineffective. The general nature of the mechanism is supported by the fact that no rearrangement to the *meta*-position was detected, and by the fact that in the presence of a scavenger for bromine (acetone was found to be suitable), debromination could be made the predominant reaction.

Similar aromatic rearrangements catalysed by aluminium bromide have been observed by Kooyman and Louw (1962). A number of re-arrangements of bromosteroids probably proceed similarly (Kirk and Hartshorn, 1968), though when hydrogen and bromine are directly in competition for reaction with nucleophilic bromide ion, the former is often the more rapidly removed.

7.3.7 Temperature coefficients for bromination. Halogenations are activated processes; and a certain amount of use has been made of temperature coefficients of the rates of chlorination. Bradfield and Jones (1941) have reviewed some of these studies, and have shown that for phenolic ethers and anilides the changes in rate are to a major degree reflected in the measured Arrhenius activation energies. This conclusion has been extended and confirmed for the chlorination of a number of hydrocarbons of varying reactivity (de la Mare and Lomas, 1965). The reactions are 'slow', in that they have negative entropies of activation; the transition state must be regarded as more restricted in its modes of motion than the initial state, as is often the case for bimolecular reactions of neutral molecules.

   Although the temperature coefficients of the rates of some bromine additions have been reported (Williams, 1941) these measurements have not been used extensively for mechanistic discussion, because the multi-stage nature of the reaction makes interpretation uncertain. Because of the formation of complexes in pre-equilibrium, the values of the measured Arrhenius parameters do not have a simple meaning (see also §1.13). An example is found in the third-order bromination of allyl benzoate in chlorobenzene (de la Mare, Scott and Robertson, 1945), which has a temperature coefficient of unity, and hence an apparent energy of activation of zero. This, it is believed, arises because the heat of dissociation of the complex (either [Olefin, Br_2] or [Br_4]) formed in pre-equilibrium is approximately equal to the activation energy for the further reaction of the complex.

7.4   The intermediates in reactions initiated by electrophilic bromine

7.4.1   Introduction. The kinetic investigations described above indicate a general similarity between molecular bromination and molecular chlorination; the former reaction becomes slightly more complicated through the fact that bromine becomes polyvalent more readily, and is more easily released as an electrophile. The general scheme given in chapter 3 (fig. 3.1, §3.6) can be used for discussion of the partly successive and partly concomitant intermediates of five different kinds: isomeric molecular complexes, isomeric σ-complexes, isomeric ion-pairs, isomeric carbocations and rearranged carbocations. Any of these could be responsible in various ways for the formation of primary products of substitution or addition, and in their turn could be responsible for the further formation of derived products.

7.4.2 Isomeric molecular complexes involving molecular bromine. The occurrence of molecular complexes between bromine and unsaturated compounds has been recognised, particularly in aprotic solvents. They are most easily investigated for situations which do not lead to substitution; thus Buckles *et al.* (1952, 1958) have shown by studies of conductivity and absorption spectra that tetrakis(p-methoxyphenyl)ethylene reacts with bromine in ethylene chloride through several stages, the first of which probably involves the formation of a loose complex of stoicheiometry [$Ar_2C:CAr_2$, Br_2]. In such a complex, for which several isomeric forms of different geometry can be written, the double-bond system would be expected to be modified only to a minor extent by complex formation. In principle, there are several ways in which the formation of such complexes might influence the details of the reaction path. First, a molecular complex might involve a site in the organic substrate remote from the position at which reaction is ultimately to be completed; under these circumstances, reduced reactivity might be found. Secondly, a particular orientation of the reacting molecules might be favoured by formation of a complex, so that particular routes to products might be promoted. Thirdly, a molecular complex might be formed in a pre-equilibrium step essential to reaction because otherwise the reagents could not be brought together fast enough. Fourthly, a molecular complex might itself be an effective electrophile. In practice, however, it is not yet clear whether molecular complexes play an essential role in bromination by molecular bromine; a possible example for bromination by hypobromous acid was noted in §7.2.

7.4.3 Isomeric σ-complexes involving molecular bromine. For chlorination, the intervention of σ-complexes involving covalent bonding between the whole chlorine molecule and the unsaturated compound appears to result in modification of the reaction path in a number of ways: partly by enhancing the reactivity of particular substrates, thus giving rates of chlorination which correlate poorly with theoretical parameters; and partly (again with particular substrates) by giving high proportions of *syn*-addition through a geometry particularly favourable for capture of the departing nucleophile by the carbocationic centre.

For bromination, the formation of these complexes seems to be manifested rather differently, being reflected more by the appearance of kinetic forms of high order in bromine and less by the stereochemistry of the reaction products. It appears that the tendency of the bromine complexes to lose tribromide ion, and of the resulting carbocation to allow the halogen to bridge between two carbocationic centres, are features which help to prevent *syn*-capture of bromide ion. The fact that bromide ion is still present in the rate-determining stage of many brominations, however, is shown by the good correlations between the rates of chlorination and bromination of aromatic and olefinic substrates; and, still more directly, by the fact that molecular bromination is much more subject to steric hindrance than bromination involving 'positive bromine', as was noted in §7.2.

Evidence concerning the structures of these complexes is indirect and inferential in character; the best way of representing the covalent bonding, and the various possible geometries, was discussed in chapter 2, and is related to the problem of the structures of the trihalide ions. The writer has generally favoured formal representations such as (7.12); the no-bonded structure (7.13) contributes to a resonance hybrid of the two, so that (7.14), intermediate between the two, gives a more accurate picture of the charge distribution. If such a complex could be examined by physical methods, its spectroscopic properties would be expected to show that both the unsaturated system and the bromine molecule had become profoundly modified, so that features related both to the formation of the carbocationic centre and of the dicovalency of a halogen atom would appear.

$$\begin{array}{ccc} \overset{+}{>}\!C\!-\!C\!< & \overset{+}{>}\!C\!-\!C\!< & \overset{+}{>}\!C\!-\!C\!< \\ |\quad\; & |\quad\; & |\quad\; \\ Br\!-\!Br^{-} & Br^{-}\; Br & Br\!-\!Br \\ & & \underline{\quad\quad} \\ (7.12) & (7.13) & (7.14) \end{array}$$

7.4.4 Ion-pairs as intermediates in electrophilic brominations; and a note on brominations in aprotic solvents. The dissociation of such covalently bound intermediates as have been discussed in the previous section gives carbocations and halide ions as products of full heterolysis. In principle, this reaction can involve further intermediates, namely ion-pairs or other more complex ionic aggregates, in which the covalent bond between the halogen atoms has been broken but the intermediate behaves kinetically as an independent species requiring activation for its further reactions. The spectroscopic properties of an ion-pair formed in this way would be very nearly that of the free carbocation and the free halide ion. Its chemical properties could, however, be special: first, because of the proximity of the counter-ion, and secondly, because facile exchange with other ions in the medium might have become important.

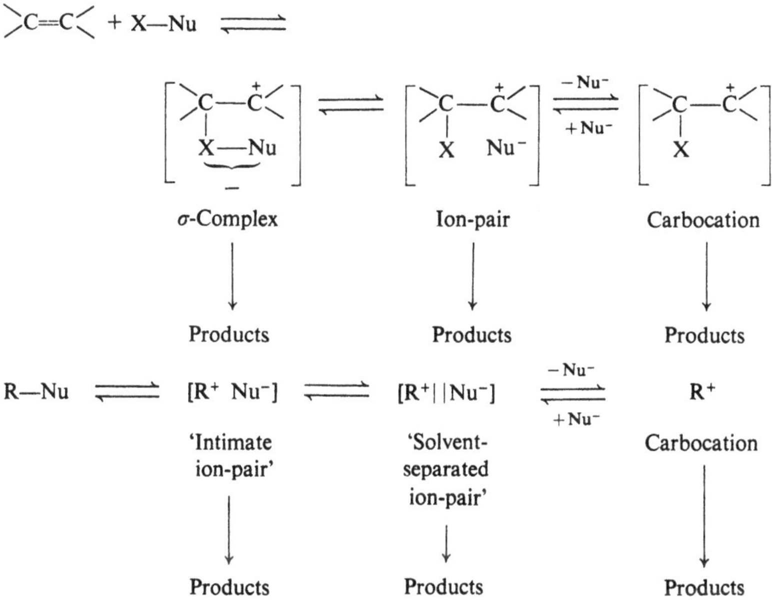

Fig. 7.3. Comparison of proposed reaction paths for electrophilic halogenations and nucleophilic replacements.

For discussions of more elaborate possibilities, involving a still greater number of potentially product-determining intermediates in nucleophilic substitution, see de la Mare and Swedlund (1973). It can be seen that the present writer regards 'intimate ion-pairs' of this reaction as analogous with the σ-complexes of aromatic substitution, a view which stresses the residual covalent bonding in such intermediates.

It was noted in chapter 3 (§3.5.9) that the importance of the successive formation of intermediates of this general kind has been recognised through studies of rates, products, and stereochemistry of nucleophilic replacements and rearrangements; comparison of the two sequences shown in fig. 7.3 indicates the analogies between behaviour in the two areas. Although there can be little doubt that the phenomena important in the chemistry determined by carbocationic intermediates in nucleophilic replacements must be reflected also in the chemistry of electrophilic processes in which similar intermediates are formed, there have been few studies which establish the participation of ion-pairs as distinct from other intermediates. The rearrangement, (7.19), of the dienone shown in structure **(7.15)** is relevant, however, since it represents a late stage in the bromination of 2,6-di-t-butylphenol. The rate of this acid-catalysed reaction, and of the accompanying reaction catalysed by bromide ions, is best interpreted in terms of an equilibrium between ion-pairs, (7.20), the spectroscopic properties of which approximate closely to those of the

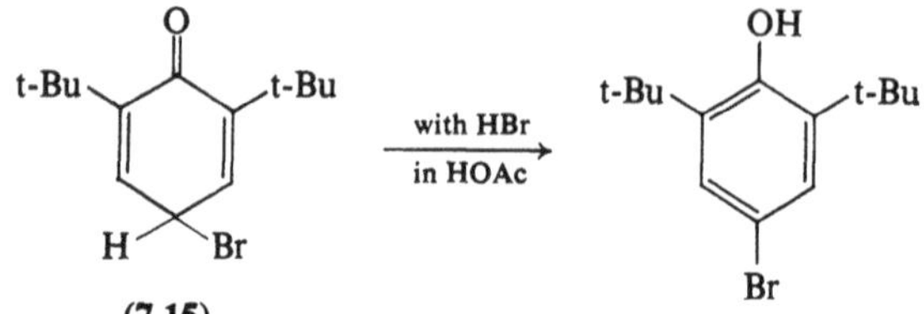

$$\text{(7.19)}$$

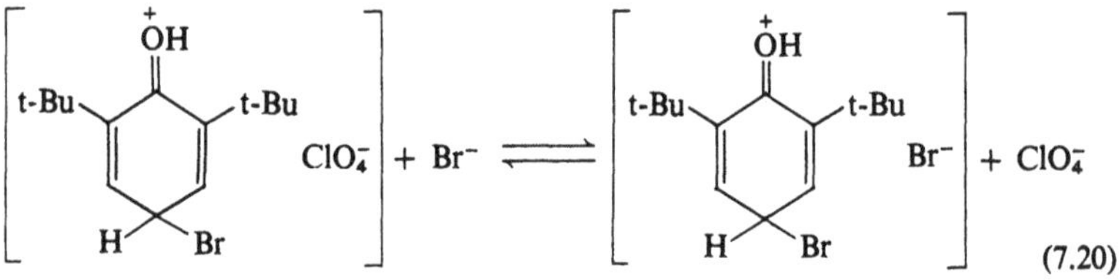

$$\text{(7.20)}$$

free ions (de la Mare and Singh, 1972). Since debromination accompanying the rearrangement is more important through the bromide-containing ion-pair, (7.17), than through the perchlorate-containing ion-pair, (7.16), it is clear that the overall course of the reaction can be altered by ion-pairing.

The less the dielectric constant of the solvent in which halogenation is conducted, the more likely it is that ion-pairing will be important. In practice, each solvent or group of solvents introduces its own particular features and complexities, and for only a few of them have wide-ranging kinetic and mechanistic studies been made. For reactions in the most non-polar solvents (saturated hydrocarbons; unsaturated hydrocarbons including benzene; carbon tetrachloride), homolytic and heterogenous processes are sometimes difficult to exclude; but heterolytic reactions, if they can be realised, give products uncontaminated by capture of the carbocation by the solvent. Solvents of this kind, in which halogenations are often slow, are most suitable for reactions of olefinic or of aromatic compounds in which the system is activated by electron-releasing groups or with catalysts of the type which promote the heterolytic fission of the interhalogen bond. They have occasionally been used also for the investigation of the properties of carbocationic ion-pairs. Thus Huyser and Kim (1968) have recorded that the reaction of 1-mesityl-1-methyl-ethylene, in which the potential carbocationic centre is heavily congested, reacts with bromine in n-pentane at low temperatures to give a highly coloured compound, formulated as the ion-pair shown in (7.21), which decomposes on being warmed, or in carbon tetrachloride, with loss of hydrogen bromide.

$$\text{Ar}.\text{C(Me)}:\text{CH}_2 \xrightarrow[\substack{\text{in pentane} \\ \text{at} -60°\text{C}}]{+\text{Br}_2} [\text{Ar}.\overset{+}{\text{C}}\text{(Me)}.\text{CH}_2\text{Br}, \text{Br}^-] \xrightarrow{-\text{HBr}} \begin{cases} \text{Ar}.\text{C(:CH}_2).\text{CH}_2\text{Br} \\ \text{Ar}.\text{C(Me)}:\text{CHBr} \end{cases} \quad (7.21)$$

Methylene chloride, chloroform and chlorobenzene are examples of solvents which, probably because of their dipolar character, are more favourable for the realisation of smooth heterolytic reactions involving halogen molecules, and are often useful preparatively.

Dipolar aprotic solvents such as nitromethane, dimethyl formamide, and dimethyl sulphoxide promote relatively fast heterolytic chlorinations and brominations. Aromatic substitutions in nitromethane have been examined kinetically (Illuminati and Marino, 1956); they resemble the corresponding reactions in acetic acid, but are faster, and have a somewhat lower ρ-value ($\rho^+ = -8.6$ in MeNO_2, -12 in HOAc). It is probable that reactions in ethers such as dioxan or in mixed solvents containing them are similar; but in all these oxygen-containing dipolar media, the specific chemical intervention of the solvent in the reaction path needs to be considered, and this can take two forms: (*a*) by the formation of a new reagent, effective both for additions and for substitutions; and (*b*) for additions, by intervention as a nucleophile and capture of the carbocationic intermediate.

The potential consequences of both of these possibilities were noted for chlorination in chapters 5 and 6. For bromine, Dalton, Smith and Jones (1970) have noted that olefinic compounds frequently give products of *anti*-addition with attachment of the oxygen function to the double bond: (7.22). More complex sequences are, however, probably available

$$\text{Ph}.\text{CH}:\text{CH}.\text{R} \xrightarrow[-\text{Br}^-]{+\text{Br}_2, +\text{CH}(=\text{O}).\text{NMe}_2} \text{Ph}.\text{CH(O}.\text{CH}:\overset{+}{\text{N}}\text{Me}_2).\text{CH(R)}.\text{Br} \quad (7.22)$$

in other similar solvents. Barili *et al.* (1972) have discussed the influence of solvent and reagent on the steric course of addition of bromine to substituted cyclohexenes. The kinetics of bromination by dioxan dibromide in benzene as solvent have been reviewed by Taylor (1972); and Fieser and Fieser (1967) review the synthetical use of this and of other brominating agents, including pyridine hydroperbromide ($\text{C}_5\text{H}_5\text{N}.\text{HBr}_3$).

7.4.5   Carbocationic intermediates formed in electrophilic brominations.

After the formation of a bromine-containing carbocationic intermediate, whatever its source may be, its subsequent fate can vary greatly both with its structure and with the medium in which it has been formed. Thus Wizinger and co-workers (Pfeiffer and Wizinger, 1928) showed that stable carbocations can be identified in the reactions of 1,1-diarylethylenes with bromine; salts of the same cation were formed by the two pathways shown in (7.23) and (7.24). These workers recognised and

$$(p\text{-MeO}.C_6H_4)_2C\text{:}CHBr + HClO_4 \rightleftharpoons [(p\text{-MeO}.C_6H_4)_2C^+.CH_2Br][ClO_4^-]$$
(7.23)

$$(p\text{-MeO}.C_6H_4)_2C\text{:}CH_2 + 2Br_2 \rightleftharpoons [(p\text{-MeO}.C_6H_4)_2C^+.CH_2Br][Br_3^-]$$
(7.24)

discussed the significance of their findings in the much wider context of the theory of aromatic substitution; for this reason, their papers form an early and important landmark in the history of mechanistic organic chemistry. Similar carbocationic intermediates (e.g. structure **(7.18)**) have recently been characterised through the ^{1}H-n.m.r. spectra of products of the reaction of 1,3,5-tris(N,N-dialkylamino)benzenes with bromine in chloroform at -60 °C (Menzel and Effenberger, 1972). The intermediates $[R_2C(Br).CH\text{:}NR_2^+, Br^-]$ involved in the reactions of enamines ($\alpha\beta$-unsaturated amines, $R_2C\text{:}CH.NR_2'$ (Szmuszkovicz, 1963)) are also related.

All these intermediates enjoy their thermodynamic stability through conjugative electron release with consequent delocalisation of the carbocationic charge; they derive little, if any, stabilisation from bridging by neighbouring bromine across a double bond. Alkyl-substituted olefins give stable carbocations only under rather special conditions, at low temperatures in highly acidic solvents, when the formation of bridged carbocations has been shown by the use of ^{1}H-n.m.r. spectroscopy (Olah and Bollinger, 1968).

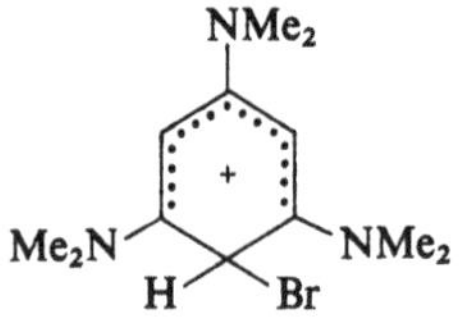

(7.18)

Normal methods of bromination of alkyl-substituted ethylenes lead towards such structures also; but they are not necessarily the first formed nor even the most thermodynamically stable cationic products. The intermediates which are produced also are often so reactive that they do not have time to reach their thermodynamically favoured geometry

TABLE 7.3 *Percentages of Markownikoff-oriented product for additions of some electrophilic reagents in water*

Electrophilic addendum	Substrate			
	$CH_3.CH:CH_2$	$HOCH_2.CH:CH_2$	$ClCH_2.CH:CH_2$	$BrCH_2.CH:CH_2$
ClOH	91	73	31	30
BrOH	79	66	27	21
BrCl	54	36	23	22

before undergoing further reaction. In this, they resemble their chlorine-containing analogues; but even so, the interaction between entering bromine and the carbocationic centre is sufficiently well developed to interfere more powerfully with other modes of possible reaction. Thus Markownikoff orientation of addition is still significant in the addition of hypobromous acid to derivatives of propylene, a result which is consistent also with the better fit of the 'unsymmetrical' than of the 'symmetrical' model to the rates of addition of bromine to alkylethylenes (table 7.2). This orientation is, however, less marked than for addition of hypochlorous acid, as is shown by the comparisons in table 7.3; and the product ratios with different nucleophiles suggest sequences involving more than one intermediate.

Similarly, the equilibration of the halogen substituents in the intermediate $[X.CH_2.CH:CH_2, X^+]$ is more marked for $X = Br$ than for $X = Cl$, but is still quite incomplete (Clarke and Williams, 1966). Furthermore, bromination of t-butylethylene in methanol gives the non-rearranged dibromide, together with the bromo-ether, (7.25), of anti-Markownikoff

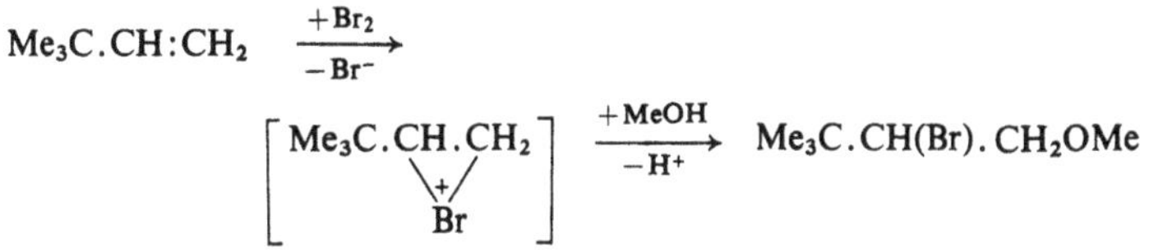

$$Me_3C.CH:CH_2 \xrightarrow[-Br^-]{+Br_2}$$

$$\left[\begin{array}{c} Me_3C.CH.CH_2 \\ \diagdown \overset{+}{Br} \diagup \end{array} \right] \xrightarrow[-H^+]{+MeOH} Me_3C.CH(Br).CH_2OMe \quad (7.25)$$

$$(7.19)$$

orientation, apparently because in the intermediate (7.19) the 2-carbon atom is so congested that the nucleophilic solvent is captured at the 1-position exclusively. A similar situation holds for addition to a number of cyclic olefinic compounds, including some important steroids, of which cholesterol is an example: (7.26) (Ziegler and Shabica, 1952).

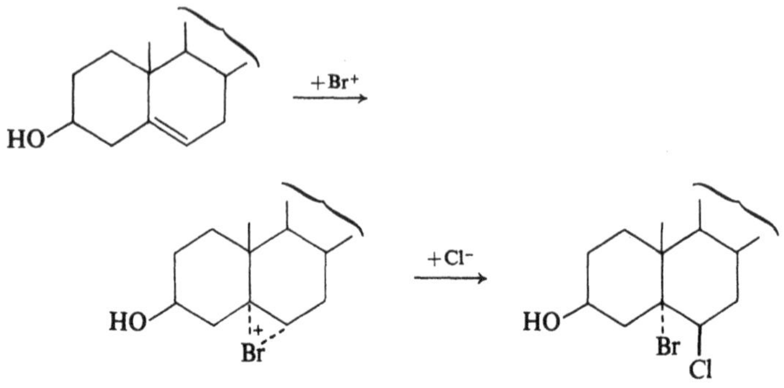

$$(7.26)$$

Additions of this kind usually proceed with *anti*-stereochemistry (Fahey, 1968), expected for reactions involving significant interaction between the entering halogen and both of the unsaturated centres. As conjugative electron release to the double bond is made more important, the carbon atom to which the electron-releasing substituent is attached more readily sustains the carbocationic charge, and has less need for additional stabilisation by neighbouring-group interaction. Consequently the observed reactions become less stereospecific, and evidence for reaction through isomerised species becomes more apparent. Fahey and Schneider (1968) have reviewed reactions of substituted styrenes which can be interpreted in this way. They examined the reactions of *cis*- and *trans*-1-phenylpropene and of *trans*-1-(*p*-methoxyphenyl)propene with bromine under conditions of kinetic control, free from disturbances from homolytic processes. The results, given in table 7.4, establish

TABLE 7.4   *Stereochemistry of addition of bromine to some 1-aryl-propenes*

		Composition of dibromides; proportion formed by:	
Olefinic compound	Solvent	*syn*-addition	*anti*-addition
trans-1-(*p*-methoxy-phenyl)prop-1-ene	CCl$_4$	0.37 (*threo-*)	0.63 (*erythro-*)
trans-1-phenyl-prop-1-ene	CCl$_4$	0.12 (*threo-*)	0.88 (*erythro-*)
trans-1-phenyl-prop-1-ene	CDCl$_3$	0.16 (*threo-*)	0.84 (*erythro-*)
cis-1-phenyl-prop-1-ene	CCl$_4$	0.17 (*erythro-*)	0.83 (*threo-*)
cis-1-phenyl-prop-1-ene	CDCl$_3$	0.22 (*erythro-*)	0.78 (*threo-*)
cis-1-phenyl-prop-1-ene	CH$_2$Cl$_2$	0.26 (*erythro-*)	0.74 (*threo-*)

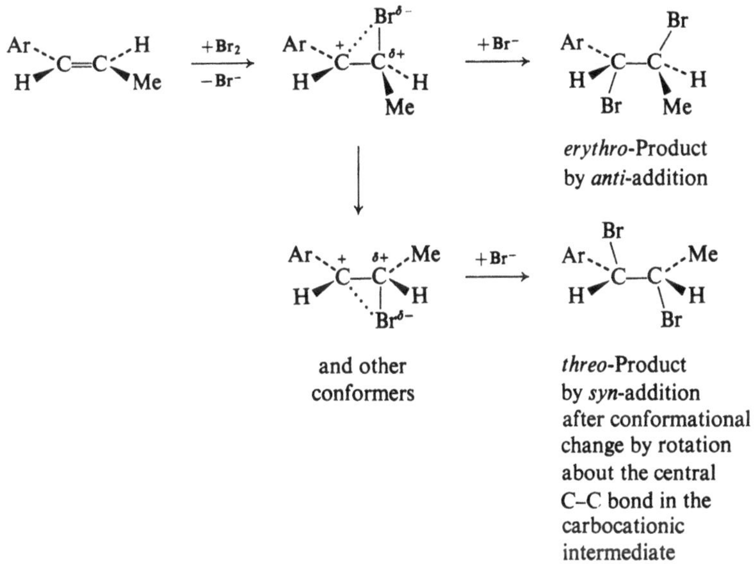

Fig. 7.4. Proposed reaction path for addition of bromine to *trans*-1-arylpropenes.

that the reactions give predominantly but not exclusively *anti*-addition in proportion which decreases as the polarity of the solvent is increased, and decreases also as the degree of electron release to the double bond is increased. The results are regarded as interpreted best by the sequence given in fig. 7.4, in which *syn*-addition becomes determined at a late stage in the reaction path; the corresponding additions of chlorine and of fluorine involve intermediates in which *syn*-addition is determined earlier.

Evidence for partial equilibration between bromine-containing carbocationic intermediates resulting from competing attack on two faces of a cyclic unsaturated system has been adduced recently by Parrilli *et al.* (1973).

7.4.6 Competition for the carbocationic centre. The carbocationic intermediates formed in bromination can often be intercepted by nucleophiles present in the reaction medium; for example (Williams, 1941), brominations in the presence of chloride ion lead to the formation of bromochlorides. Internal neighbouring groups can likewise be effective in providing the nucleophilic centre. Thus the bromination of 1,1-dimethylallyl alcohol by hypobromous acid actually takes the course shown in (7.27). Similar ring closure involving the neighbouring benz-

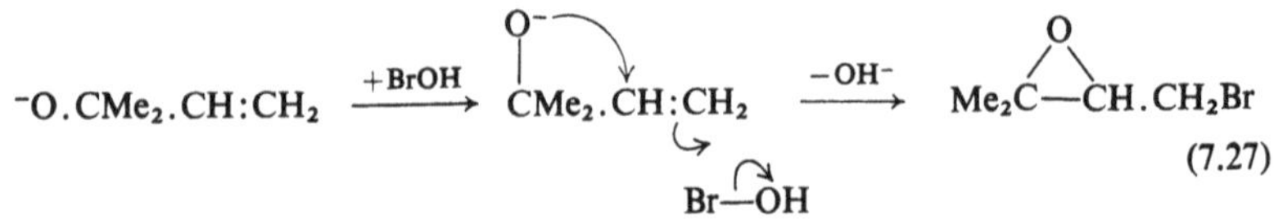

$$^-O.CMe_2.CH{:}CH_2 \xrightarrow{+BrOH} \overset{\overset{\displaystyle O^-}{|}}{CMe_2.CH{:}CH_2} \xrightarrow{-OH^-} Me_2C{-}CH.CH_2Br \tag{7.27}$$

amido-group instead of the alkoxide function has been used for stereo-specific syntheses of trisubstituted derivatives of cyclohexane through the sequence shown in (7.28).

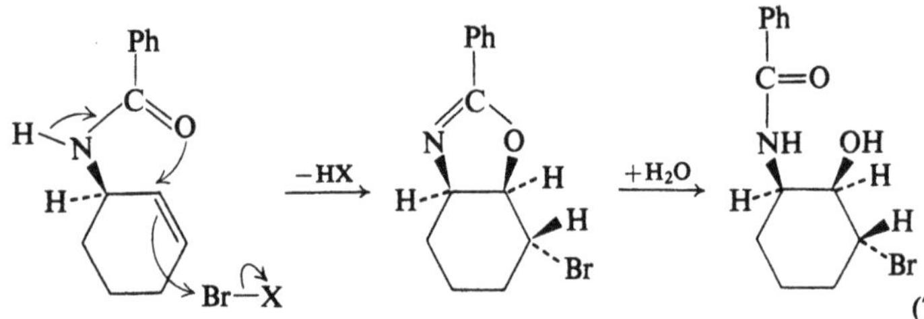

$$\tag{7.28}$$

Kinetic evidence that the neighbouring group can enhance the rate of reaction, and so contribute to the rate-determining stage of addition, has been provided by Williams, Bienvenue-Goetz and Dubois (1969) through measurements of the rates of bromination of a series of compounds $CH_2{:}CH.(CH_2)_n.OH$ ($n = 1{-}4$). These show a maximum at $n = 3$, instead of the regular increase expected in the absence of rate enhancement by neighbouring-group interaction.

7.4.7 Rearrangements of the carbocationic intermediate. Examples of Wagner–Meerwein rearrangement within a carbocationic intermediate produced in the course of electrophilic bromination are well known, particularly among bicyclic olefinic compounds. Some of these reactions have been reviewed by Fahey (1968) (see also §8.2). Among the interesting related examples found among the reactions of aromatic compounds is the bromination of benzcyclobutene with accompanying ring opening: (7.29) (Lloyd and Ongley, 1965). Double-bond rearrangements are, of

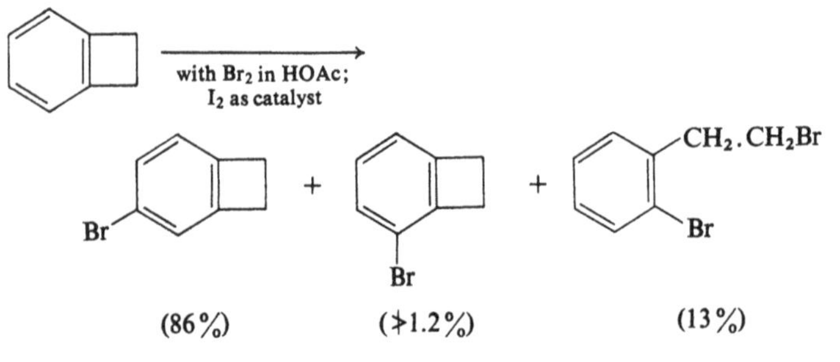

$$\tag{7.29}$$

course, as common in brominations as they are in chlorinations; they
have been reviewed by de la Mare (1963). For additions to conjugated
systems, terminal attack on the double bond gives an allylic cation
in which the positive charge is distributed by resonance, and so
1,3-butadiene (CH_2:CH.CH:CH_2) gives *trans*-1,4-dibromobut-2-ene
(Br.CH_2.CH:CH.CH_2Br) by 1,4-addition with rearrangement,
together with 3,4-dibromobut-1-ene (Br.CH_2.CH(Br).CH:CH_2) by
1,2-addition, the former being formed in slight preponderance.

Addition of bromine to vinylacetylene in chloroform provides a
particularly striking example of a kinetically controlled addition. The
products of addition to the double or triple bond form only minor
components of the reaction mixture; the major product is the allene,
Br.CH_2.CH:C:CHBr, formed by 1,4-addition with rearrangement.
This compound, though thermodynamically much less stable than its
isomer CH_2:CH.C(Br):CHBr, is formed more easily because electrophilic attack on the triple bond gives a transition state of geometry
approximating more nearly to that of the allenic (**7.20**) than of the 1,3-
dienic form (**7.21**). Because of their different geometries, structures
(**7.20**) and (**7.21**) do not form an effective resonance hybrid, so an
intermediate is formed having its charge temporarily localised as in the
former structure, and reaction is completed by nucleophilic capture to
give the allenic product.

(7.20)   (7.21)

The stereochemistry of 1,4-addition has not been investigated extensively; *cis*-1,4-addition has been reported as the major process involved
in the reaction of bromine with cyclopentadiene.

7.5   Reactions initiated by brominating species other than molecular bromine

7.5.1   Bromine fluoride and bromine chloride.

The elements of bromine
fluoride can be added to olefinic compounds by using *N*-bromoacetamide
in anhydrous hydrogen fluoride containing a little diethyl ether. The
reaction has been reviewed by Boguslavskaya (1972); whether the
bromine fluoride molecule is the effective electrophile does not seem to

be known. Its disproportionation to $[FBrF^+]F^-$ would make it tend to be
a fluorinating, rather than a brominating agent. Molecular bromine
chloride, on the other hand, is itself able to provide electrophilic bromine
both for additions and for substitutions. It can be formed from bromine
and chlorine, or from other sources of electrophilic halogen and the
appropriate halide ion, and is intermediate between molecular chlorine
and molecular bromine in reactivity. The course of the reaction, which
follows the expected mechanistic details, has been discussed by Hageman
and Havinga (1966).

7.5.2 Compounds containing bromine–oxygen bonds. There are a number
of oxygen-containing species which have been shown to be kinetically
effective in brominations, as has been reviewed by Taylor (1972). The
rates and products of bromination of biphenyl by hypobromous acid
in 75 per cent acetic acid indicate that there is a component of the rate
which follows the kinetic form

$$-d[BrOH]/dt = k[ArH][BrOAc]$$

and involves molecular bromine acetate, promoting partial rate factors
similar to those for molecular bromination rather than to those for
'positive bromination'.

Bromine acetate has been held to be concerned also in brominations
by, and rearrangements of, N-bromoanilides in aprotic solvents catalysed
by acetic acid. Under some circumstances the rate is found to be independent of the concentration of added substrate, so that the rate-
determining stage of the reaction is that of the formation of bromine
acetate (7.30) (Israel, Tuck and Soper, 1945). The suggestion that a more

$$Ar.N(Br)Ac + HOAc \longrightarrow BrOAc + Ar.NHAc \qquad (7.30)$$

complex mode of reaction is possible involving a cyclic transition state
(**7.22**) and a different kinetic form has recently been made by Scott and
Martin (1965) on the basis that the ratio of *ortho*- to *para*-bromination
of anisole changes with change in the catalysing acid.

Beebe and Wolfe (1970) have found that bromine acetate in carbon
tetrachloride can be used for the N-bromination of amides and imides; it
is proposed that a cyclic transition state (**7.23**) is adopted here also.

t-Butyl hypobromite has also found some utility for additions of
bromine, as has been recorded by Kergomard and his co-workers (Dauphin, Kergomard and Scarset, 1973, and papers therein cited). Its use

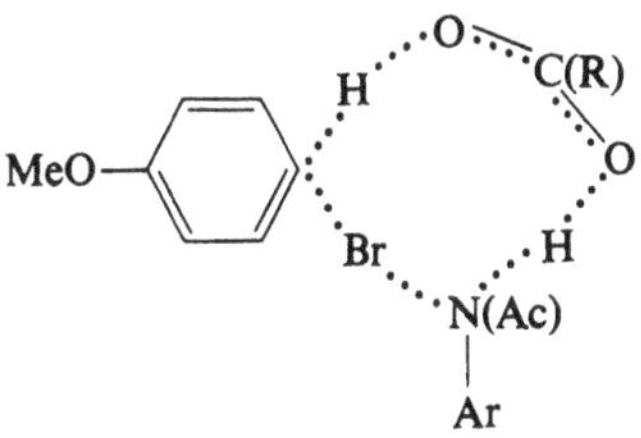

(7.22) 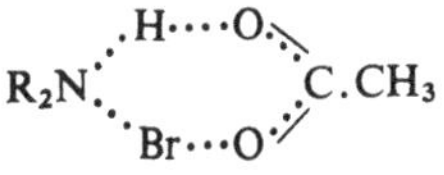   (7.23)

for some aromatic substitutions has been discussed by Mach and Bunnett (1974), and is discussed in chapter 9 in relation to other base-catalysed brominations involving carbanions.

7.5.3   Compounds containing bromine–nitrogen bonds. N-Bromosuccinimide and related compounds are very effective brominating agents; which, like their chloro-analogues, can be used to supply electrophilic halogen without the development of mineral acid. They are, however, also very effective for homolytic brominations, so that conditions conducive to these reactions need to be avoided if reaction through the heterolytic pathway is required.

The kinetics of these reactions have not been very extensively studied. Bodrikov, Bronnikova and Okrokova (1973) and co-workers have investigated the bromination of a series of 2-arylpropenes by N-bromosuccinimide in acetic acid. The products were shown to be bromoacetates of the expected orientation: (7.31). A second-order rate law was followed,

$$R.C_6H_4.C(Me):CH_2 \xrightarrow[-C_4H_5O_2N]{+C_4H_4O_2NBr,\ +HOAc} R.C_6H_4.C(Me)(OAc).CH_2Br$$
(7.31)

and the response in rate with change in substituent R correlated well with the electrophilic substituent constants σ^+, with $\rho^+ = -5.69$; the corresponding value for molecular bromination is significantly less (-4.2).

Although these results do not give complete proof that the N-bromosuccinimide molecule is itself a kinetic entity involved in attack on the unsaturated compound, the high value of ρ^+ is probably more consistent with this proposal than with the alternative possibility that bromine acetate, or the protonated forms of either of these reagents, could be the effective reagent. Comparison of the product ratios obtained in the reactions of 3-t-butylcyclohexene with various reagents leads to a similar conclusion (Bellucci *et al.*, 1972).

The use of N-bromosuccinimide in dimethyl sulphoxide, with capture of the carbocationic intermediate by the nucleophilic solvent, is described by Dalton *et al.* (1968, 1971).

A number of more complicated brominations can be effected by the use of N-bromosuccinimide and its analogues, and these indicate that a particularly selective brominating species is in operation. Thus whereas the bromination of 3-acetoxycholesta-3,5-diene by molecular bromine is complicated by the concurrence of addition processes, the corresponding bromination by N-bromosuccinimide smoothly gives 6β-bromocholest-4-en-3-one, the product of bromodeacylation with rearrangement, (7.32), either in acetic acid or in chloroform (Reich and Lardon, 1946;

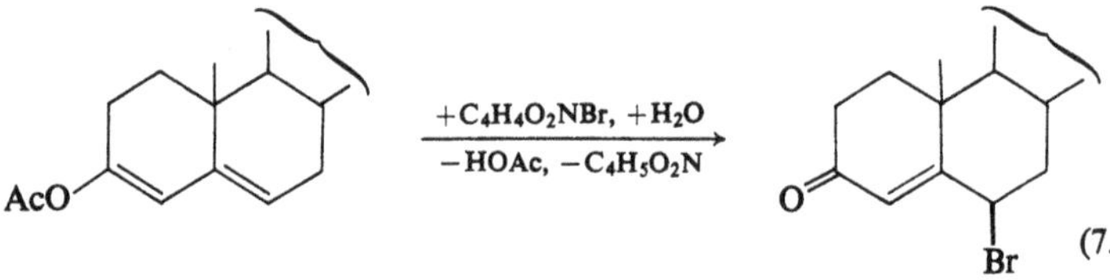

(7.32)

de la Mare and Hannan, 1973). Whether a synchronous or a multi-stage reaction is under observation is not known.

Additions with homoconjugative rearrangement, of the type shown in (7.33), can also be initiated by N-bromosuccinimide (Winstein and Shatavsky, 1956).

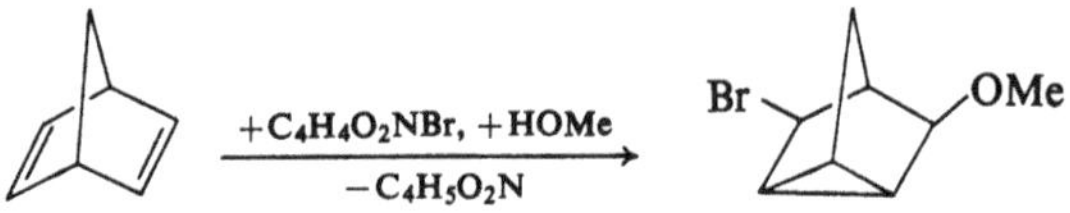

(7.33)

Heterolytic aromatic substitutions can also be effected by using N-bromoamines or amides. Pearson, Wysong and Breder (1967) have described the selective *ortho*-bromination of phenol by N-bromo-t-butylamine in toluene at $-70\,°\text{C}$; these workers favour a reaction path involving the intermediate formation of phenyl hypobromite, which then undergoes rearrangement selectively to the *ortho*-position.

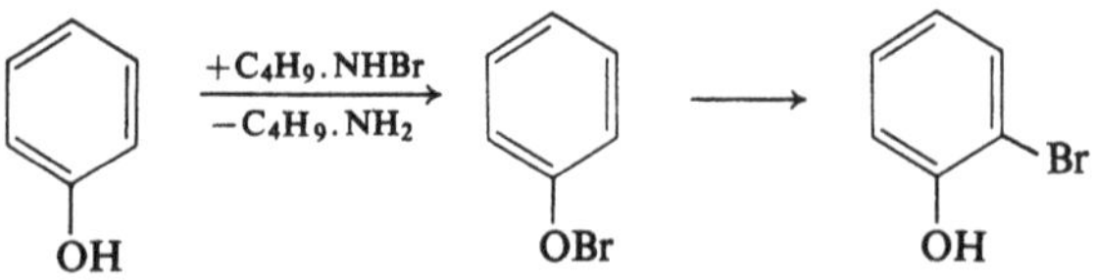

(7.34)

Bromine azide ($\text{Br}.\text{N}_3$) is a reagent which can be used for introducing a C–N bond into an organic molecule; the nature of its reactions, which resemble in stereochemistry and mechanism those of bromine chloride, have been reviewed by Hassner (1971). Equation (7.35) provides an example. Complexes of bromine with pyridine and its analogues are also effective brominating agents. Pyridine, for example, can be used in

$$Ph.CH:CH_2 + Br.N_3 \xrightarrow{\text{in } CH_2Cl_2/MeNO_2} Ph.CH(N_3).CH_2Br \quad (7.35)$$

catalytic amounts in aprotic solvents to assist the bromination of aromatic compounds, and it has often been assumed that the function of the heterocyclic compound is to provide a new source of electrophilic bromine (e.g. $C_5H_5NBr^+$). The use of pyridine hydroperbromide ($C_5H_5NH^+Br_3^-$) as a selective brominating agent has been described by Fieser and Fieser (1967); it gives products predominantly of *anti*-addition.

7.5.4   Compounds containing bromine–carbon bonds. A few compounds containing Br–C bonds have been used as electrophilic brominating agents. Recently, for example, Hallas and Hepworth (1974) have described the use of 2,4,4,6-tetrabromocyclohexa-2,5-dienone (**7.24**) for

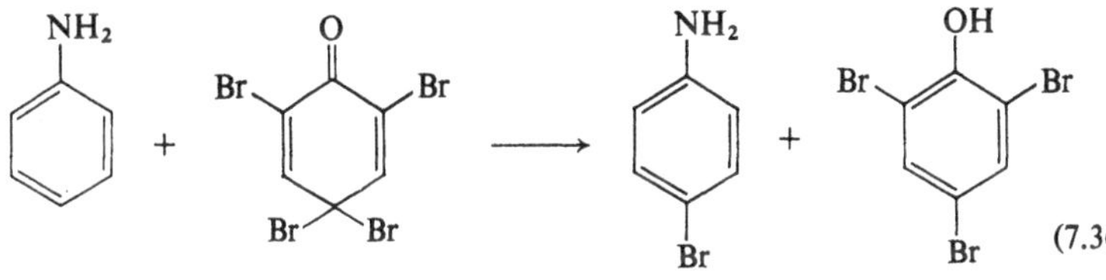

(7.36)

the monobromination of amines, (7.36), which with molecular bromine very readily give di- and poly-brominated products. It is probable (Calo *et al.*, 1974) that this reagent does not normally react directly with unsaturated compounds, but instead reacts with small concentrations of nucleophiles, Nu⁻, to provide brominating species of the type Br–Nu (e.g. BrCl, when HCl is used as catalyst; see also §§7.3.1, 7.4.4).

7.5.5   Metal halides and related compounds as reagents or catalysts for bromination. As for chlorination, metal halides can act in a number of ways as reagents or catalysts for bromination. Thallium-catalysed brominations have recently been investigated by several groups of workers. Taylor and McKillop (1970) have reported that a number of substrates, including aromatic ethers, are brominated by bromine with thallium triacetate as catalyst to give *para*-substituted derivatives nearly exclusively. They suggest that the orientation is determined by the steric bulk of the electrophile supplied as an ordered complex of stoicheiometry

[ArH, Br_2, $Tl(OAc)_3$]. Erickson and Barowsky (1971) have thrown doubt on this conclusion, however, by showing that large alkyl groups adjacent to the position *para* to the directing ether function do not divert substitution to the now less hindered *ortho*-position. Uemura *et al.* (1971) have compared the partial rate factors for bromination of toluene by thallium salts in carbon tetrachloride and other solvents at various temperatures with results for other brominations, with the results shown in structures (**7.25**)–(**7.30**). They conclude that derivatives of thallium(III) give, as

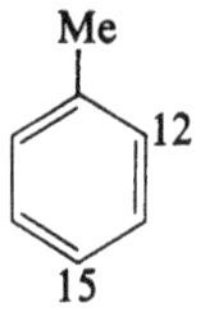

Me ⟨ring⟩ 12 15	Me ⟨ring⟩ 15 12	Me ⟨ring⟩ 76 59
$TlBr_3$, $4H_2O$ CCl_4 39 °C (**7.25**)	Br_2, $FeCl_3$ $MeNO_2$ 25 °C (**7.26**)	$BrOH_2^+$ 50 % dioxan 25 °C (**7.27**)
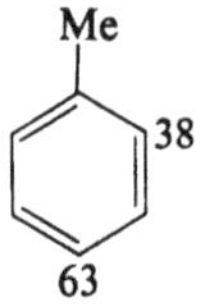 Me ⟨ring⟩ 38 63	Me ⟨ring⟩ 636 2938	Me ⟨ring⟩ 600 2420
Br_2, $Tl(OAc)_3$ CCl_4 77 °C (**7.28**)	Br_2, $TlBr$ CCl_4 77 °C (**7.29**)	Br_2 85 % HOAc 25 °C (**7.30**)

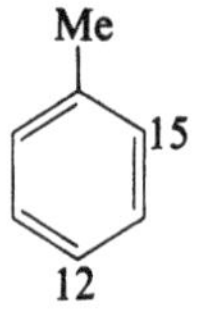

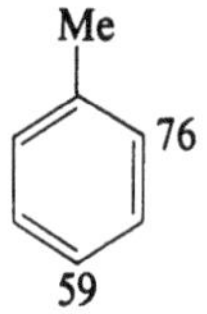

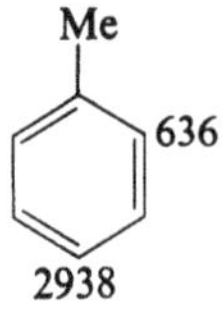

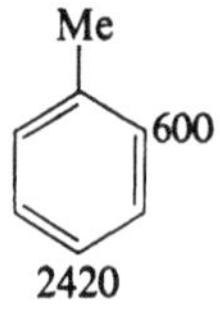

ferric chloride does, a reagent rather like $BrOH_2^+$ in its response to change in substituent and to steric effects; whereas derivatives of thallium(I) promote a reaction of molecular bromine, or provide a reagent of similar properties. In the former cases, it seems reasonable to suppose that ion-pairs (e.g. $[Br^+TlBr_2X_2^-]$ or $[Br^+FeBrCl_2^-]$) are concerned.

7.6 Secondary product-determining pathways in bromination

The carbon–bromine bond is more easily broken than the carbon–chlorine bond, irrespective of whether a homolytic process, a heterolytic process leading to Br^-, or a heterolytic process leading to Br^+ is concerned. Accordingly, later stages of the reaction path tend to be more important in bromination than in chlorination; and it becomes more easy to fail to recognise the existence and significance of earlier stages on the reaction path.

This problem becomes particularly acute in relation to the possible incursion of addition–elimination sequences leading to substitutions. An instructive example has been given recently by de la Mare and Wilson (1974). The kinetically controlled reaction (7.37) of bromine with cholest-5-en-3-one in acetic acid containing sodium acetate gives, among other

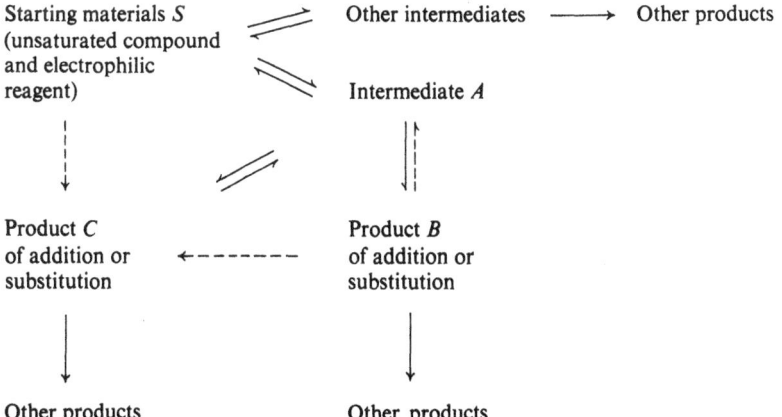

products, $5\alpha,6\beta$-dibromocholestan-3-one (**7.31**), which eliminates hydrogen bromide either heterolytically or, with great ease, homolytic-ally, to give the rearranged product of substitution, 6β-bromocholest-4-en-3-one (**7.32**). It is the homolytic reaction to which particular attention is drawn. Such a process is formally the reverse of the much-studied peroxide-catalysed addition of hydrogen bromide, and in this case is

Starting materials S (unsaturated compound and electrophilic reagent) ⇌ Other intermediates ⟶ Other products

Intermediate A

Product C of addition or substitution ←------- Product B of addition or substitution

Other products   Other products

Fig. 7.5. Direct and indirect routes to products of reaction of unsaturated compounds with electrophiles.

TABLE 7.5 *Examples of reactions which may proce*

Starting materials (*S*)	Intermediate (*A*)	Product (*B*)
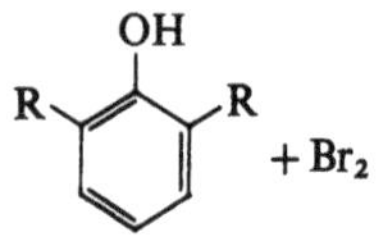	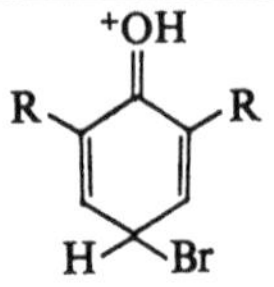	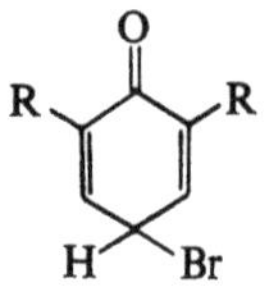
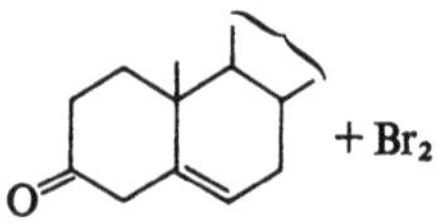	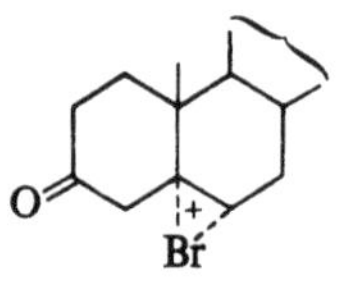	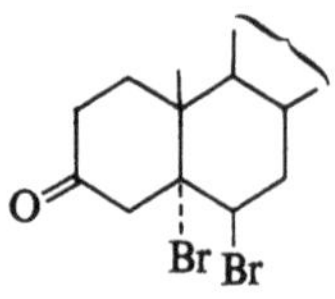
$R.CH{:}CH.CH{:}CH_2$ + Br_2	$[R.\overset{+}{\overbrace{CH.CH.CH}}.CH_2Br]$	$R.CH{:}CH.\underset{\displaystyle Br}{\underset{\vert}{CH}}.CH_2$
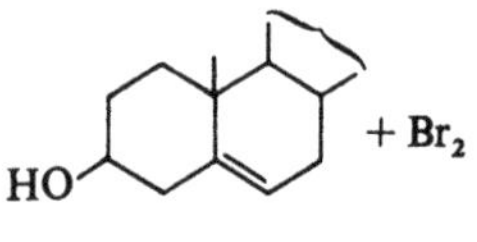	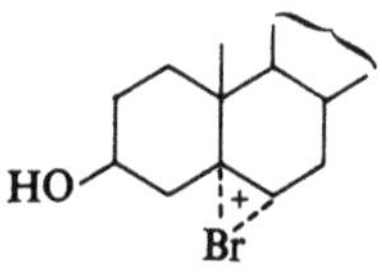	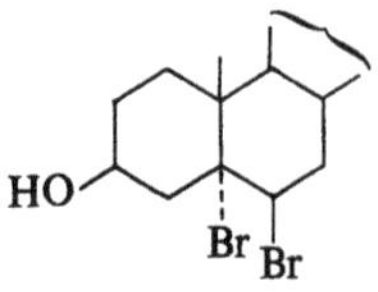
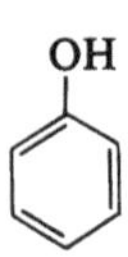 or 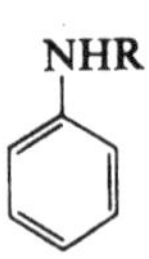	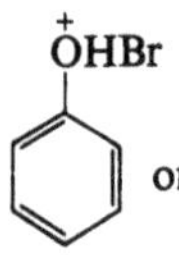 or 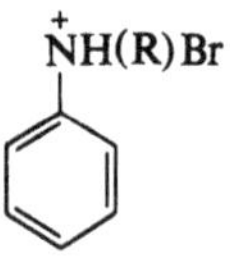	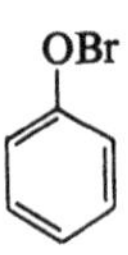 or 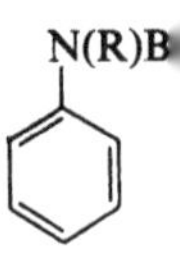

[a] The categorisations are those which the writer believes woul normally be thought to be correct formally. They do not impl that the reaction categorised is (or is not) a two- or multi-stag process, but the word 'direct' is meant to exclude the possibilit 'via *A*'. Not all the 'direct' processes are known to exist; but fo all those with an entry 'Rearr.', experimental evidence exist for an intramolecular process of electronic requirement unspecified. In all the examples given, there is evidence tha the conversion $S \rightarrow C$ can be carried out by at least tw reaction paths: $S \rightarrow A \rightarrow C$ and $S \rightarrow A \rightarrow B \rightarrow$ (not via *A*) *C*

hrough direct or indirect routes as outlined in fig. 7.5

Product (C)	Classification of routes[a]						Cross-references
	$S \rightarrow B$ 'direct'	$S \rightarrow B$ via A	$S \rightarrow C$ 'direct'	$S \rightarrow C$ via A	$B \rightarrow C$ 'direct'	$B \rightarrow C$ via A	
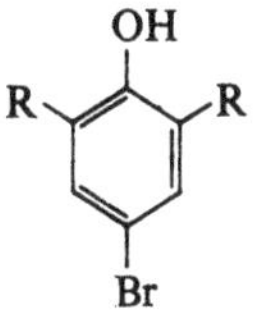 [2,6-di-R-4-bromophenol]	S_E2'	S_E2'	S_E2	S_E2	Rearr.	S_E2'	Note b; chapters 5 and 6; fig. 7.1; equations (7.6), (7.19)
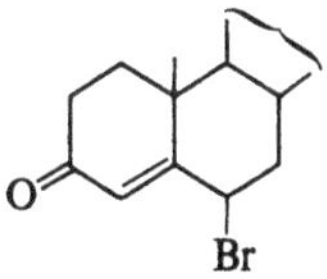 [bromo-substituted octahydronaphthalenone]	Ad	Ad	S_E2'	Ad, then E	E_H	E_1	Equation (7.36)
$R.CH.CH{:}CH.CH_2Br$ with $\underset{\mid}{Br}$	Ad	Ad	Ad'	Ad'	Rearr.	S_N1'	§7.4.7; note c
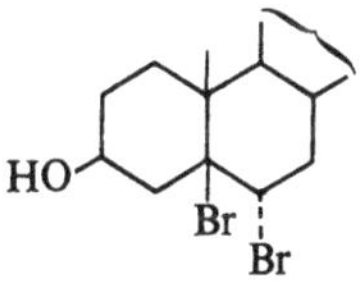 [HO-decahydronaphthalene with Br, Br]	Ad	Ad	Ad	Ad	Rearr.	S_N1'	Note d
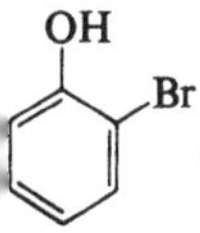 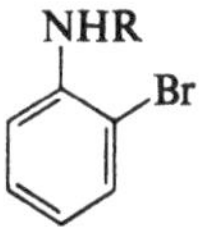 [2-bromophenol (OH, Br)] or [2-bromo-N-R-aniline (NHR, Br)]	S_E2 (at O or N)	S_E2 (at O or N)	S_E2	S_E2	Rearr.	S_E1'	Equation (7.34); (10.4) and accompanying text; note e

[a] Electrophilic rearrangements of C, and nucleophilic or electrophilic rearrangements of B, are also possible and may lead to isomeric products.

[b] For discussion of mechanisms of allylic rearrangements, see de la Mare (1963).

[c] For discussion of mechanisms of axial–equatorial rearrangement, see Grob and Winstein (1952).

[d] Ahmed and Wardell (1972) discuss a related example in relation to substitution in alkyl phenyl sulphides.

favoured thermodynamically because of steric congestion in the dibromide, (**7.31**), and of conjugation in the final product, (**7.32**).

With other brominating agents, or in the presence of a suitable nucleophile, the initial stage of bromination with capture of a nucleophile
followed by elimination of hydrogen bromide gives a 6β-substituted
product in which the substituent was not the originally electrophilic
group, as is indicated also as a possibility in (7.37).

Sequences of heterolytic followed by homolytic reactions can be
difficult to identify, since the kinetics of the reaction will be characteristic
of the former type of process, whereas the products may be typical of
either type of reaction. Such complications may be much more common
in the reactions involving unsaturated bromides than has hitherto been
recognised. In fig. 7.5, the mechanistic problem under consideration is
presented in a more general way, consistent with fig. 3.1. When a particular product C is identified from a reaction involving starting materials S,
and is consistent with reaction through an intermediate A, is C formed
directly from A or indirectly via a second product B? If C is formed from
B, does B give C via the same intermediate A, or by some other mode of
reaction? Examples, relevant particularly to brominations, are given in
table 7.5 to illustrate the widespread need to ask questions of this kind;
they apply to other electrophilic additions and substitutions also. Most
of the so-called 'unusual' or 'less common' mechanisms of aromatic
substitution, including addition–elimination, substitution with rearrangement, and 'indirect substitution', present the investigator with
such possibilities; many aromatic rearrangements also need consideration in these terms (Hughes and Ingold, 1952; de la Mare, 1971).

Note added in proof, February 1976

The claim by Schubert and Dial (1975; *J. Amer. Chem. Soc.* **97**, 3877) that the mechanism represented by (**7.17**) and (**7.18**) has a kinetic form the same as, or equivalent to,
that representing a combination of mechanisms involving one (**7.3**) and more (e.g.,
(**7.8**)) molecules of bromine is incorrect, except in the trivial circumstance that the terms
of order higher than one in bromine are negligible, and then only with further provisos.
The kinetic forms can resemble each other for a single kinetic run at a single initial
concentration of bromine, but become distinguished clearly when the initial concentration of bromine is varied over an appropriate range, as has been done by the original
and many subsequent workers (for summary, see for example Taylor (1972)).

8 The iodination of unsaturated compounds

'...the secret of the sunbeam...' (R. Browning)

8.1 Introduction

The beautiful colour of iodine vapour, and of its solutions in aprotic solvents, speaks for the fact that it is the most polarisable of the common halogens. It is also the largest; the one which most readily bears a positive charge; the one which forms the weakest bonds to carbon and the other first-row elements; and the one which has the greatest tendency to become di- or poly-ligant.

The balance of these factors produces the result that the iodine molecule is the least reactive of all the molecular halogens in effecting addition to, or substitution in, unsaturated compounds; and that the reversibility of iodination often becomes apparent in the incompleteness of reaction, or in the formation of products more nearly of thermodynamic control, or in the observation of primary kinetic isotope effects on the rates of substitution reactions.

Homolytic processes can also sometimes become evident; but, as with the addition of hydrogen iodide to unsaturated compounds, iodine atoms are self-scavenging and do not easily sustain long chain processes; so it is less easy than with bromine to alter the reaction conditions sufficiently to make free-radical reactions dominant.

Since the iodine cation is thermodynamically much more stable than the similar cations of fluorine, chlorine or bromine, it might have been thought that reactions involving I^+ with unsaturated compounds would be easy to identify. In fact, this turns out not to be the case; reactions in which electrophilic iodine is supplied by a neutral molecule are much more common, and even those for which an iodine-containing cation can be identified kinetically more commonly involve positive iodine covalently bound to a nucleophilic species, as in the cations $[I, py_2]^+$ and $[I_3]^+$.

159

Studies from the field of nucleophilic displacements have led to the conclusion that bridging from neighbouring iodine to an adjacent carbocationic centre is energetically much more favourable than for the lower halogens. It might have been thought also, therefore, that reactions through fully bridged iodonium cations, (**8.1**), would be very important in additions initiated by electrophilic iodine. Although bridging is probably significant in many such additions, their features are often better explained in terms of unsymmetrically bridged intermediates (e.g. (**8.2**)), favoured perhaps because steric congestion can then be minimised. Furthermore, it is apparent that a nucleophile is often concerned in the rate-determining step of iodination, and it can be involved either through attachment to the electrophilic iodine, (**8.3**), or in attacking the carbocationic centre, (**8.4**), or both, (**8.5**).

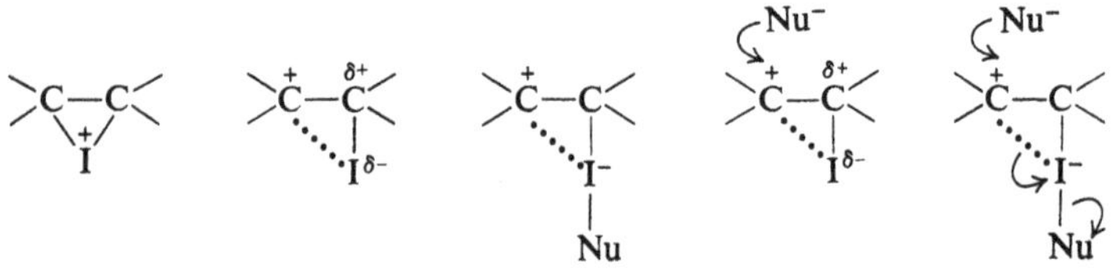

(**8.1**)   (**8.2**)   (**8.3**)   (**8.4**)   (**8.5**)

It seems likely also that molecular complexes in which carbocationic character is not fully developed are more significant in iodinations than in other halogenations. Thus study of the spectral changes occurring when iodine reacts with N,N-dimethylaniline in cyclohexane (Senkowski and Panson, 1961) led to the suggestion that the sequence shown in (8.1) is followed.

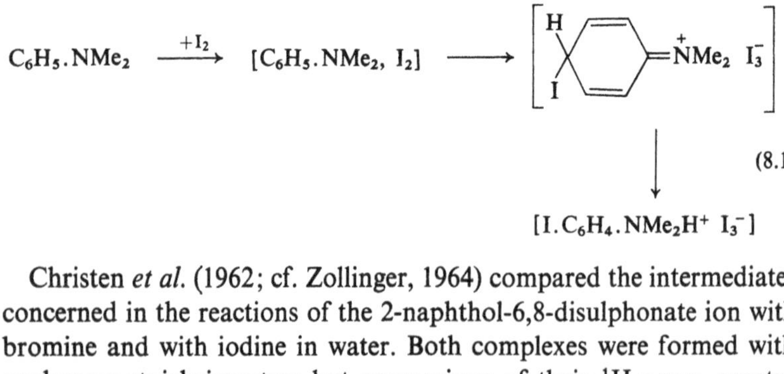

$$C_6H_5 . NMe_2 \xrightarrow{+I_2} [C_6H_5 . NMe_2, I_2] \longrightarrow \left[\text{(structure)} =\overset{+}{N}Me_2 \ I_3^- \right]$$

(8.1)

$$[I . C_6H_4 . NMe_2H^+ \ I_3^-]$$

Christen *et al.* (1962; cf. Zollinger, 1964) compared the intermediates concerned in the reactions of the 2-naphthol-6,8-disulphonate ion with bromine and with iodine in water. Both complexes were formed with analogous stoicheiometry; but comparison of their ^{1}H-n.m.r. spectra showed that the aromatic system was much more perturbed by bromine than by iodine, so that a σ-complex, (**8.6**), was concerned in the bromination and subsequently rearranged to give the normal product of substitution; whereas in the reaction with iodine, which stopped at the early stage, a much looser complex was formed. This was thought to be a π-complex, though the possibility that it was an O-iodide was not excluded.

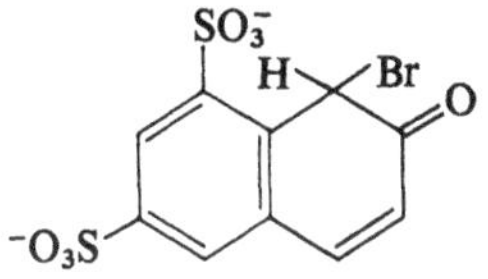

(8.6)

Other indications of the possible role of molecular complexes in iodinations come from studies of displacement of groups other than hydrogen from aromatic molecules (for review, see Taylor, 1972). Thus in the iododestannylation shown in (8.2) the effects of substituents on the rates

$$R_3Sn.C_6H_4.R' \xrightarrow[\text{in } CCl_4]{+I_2} I.C_6H_4.R' + R_3SnI \qquad (8.2)$$

of the reaction do not correlate well with those found in other typical aromatic substitutions. Similarly the iododesilylation shown in (8.3), and the iododemercuration shown in (8.4), are both unexpectedly faster

$$Ph.SiMe_3 \xrightarrow[\text{in HOAc}]{+I_2} Ph.I + Me_3SiI \qquad (8.3)$$

$$Ph.HgBr \xrightarrow[\text{in HC(:O).NMe}_2]{+I_2} Ph.I + HgIBr \qquad (8.4)$$

than the corresponding brominations and chlorinations, so it has been suggested that special four-centred transition states may be involved. Stabilisation attained in pre-formed molecular complexes possibly accounts for the availability of such structures.

For many iodinations, however, the orientational and kinetic behaviour indicates that the rates and products are determined through intermediates or transition states having much carbocationic structure, though quite various in composition and detailed structure. The following sections deal with reactions which mostly appear to be of this kind.

8.2 Iodination by molecular iodine, and by iodine chloride

Reactions of high kinetic order in the iodinating agent are well established both for additions of iodine to olefinic compounds and for aromatic substitutions by iodine chloride, solvents such as acetic acid and others of low polarity being used. Polar conditions, and specific catalysts, have been used to reduce the kinetic order with respect to the halogenating agent, just as they have for the similar reactions of molecular bromine. Thus the kinetics of the reaction of propene with iodine catalysed by peracetic acid in mixtures of acetic acid and diethyl ether were found to indicate a rate-determining attack by peracetic acid on a complex, of composition [Olefin, I_2] (Ogata and Aoki, 1966). The product, 1-iodo-2-acetoxypropane, (8.5), was that of Markownikoff orientation with the incorporation of a solvent residue. Zinc chloride behaves similarly in

$$\text{Me.CH:CH}_2 \xrightarrow[\text{MeCO}_3\text{H as catalyst}]{+I_2,\ +\text{HOAc},\ -\text{HI}} \text{Me.CH(OAc).CH}_2\text{I} \qquad (8.5)$$

catalysing aromatic substitutions by iodine chloride (Andrews and Keefer, 1956). Thallium(I) carboxylates, useful preparatively for the formation of iodocarboxylates from iodine and olefinic compounds, probably act in a related manner (Cambie *et al.*, 1974).

Additions initiated by electrophilic iodine normally proceed with *anti*-stereochemistry, and a number of cases have been cited in reviews (e.g. by Fahey, 1968). Findlay, Waters and Caserio (1971) have recently made a study of the iodination of the optically active allene, *R*(−)2,3-pentadiene, (8.7). With iodine chloride in pyridine, the optically active adduct *trans*-3-iodo(4*S*)chloropent-2-ene, (8.8) with Nu = Cl, was the

$$ (8.6) $$

(8.7) (8.8)

main product (85 per cent), the remainder being its *cis*-isomer. In methanol, with iodine chloride or with iodine, the corresponding iodo-ether, (8.8) with Nu = MeO, was formed (94 per cent). The results were compared with those for bromination under similar conditions. The formation of optically active product from optically active starting material showed that the reactions occurred through paths having well-defined *anti*-stereochemistry. Capture of a solvent residue indicated the development of carbocationic character, and the orientation showed that this develops only on the terminal carbon atom, apparently directed by the methyl group.

The products of addition of the hypohalous acids to allyl chloride (de la Mare and Bolton, 1966; see also §§6.4.3, 7.4.5) indicate that the orientation is not very strongly dependent on the effectiveness with which the electrophilic halogen bridges to the carbocationic centre; the

Markownikoff product, $Hal.CH_2.CH(OH).CH_2Cl$, is formed in proportion 0.31, 0.27 and 0.29 for $Hal = Cl$, Br and I respectively. Capture of the hydroxyl group from the solvent probably, therefore, follows very rapidly after development of the carbocationic charge.

No migration of chlorine from the 3- to the 2-position occurred in the last of these reactions, in which iodine chloride in water was used as the source of electrophilic iodine. This is expected, since attack by the electrophile on the double bond gives the entering group its maximum opportunity for bridging to the carbocationic centre, (**8.9**), and participation by neighbouring chlorine would not be expected to interfere with interaction by the better neighbouring group, iodine. In the corresponding addition of hypochlorous acid to allyl iodide, however, the intermediate, though starting out in the form of (**8.10**), would if it were long lived be converted into the isomeric, more stable form of (**8.9**),

$$\overset{\delta-}{I}\cdots$$
$$\overset{|}{\underset{\delta+}{CH_2}}\overset{\cdots}{\overset{|}{CH^+}}.CH_2Cl$$

(**8.9**)

$$\cdots\overset{\cdots}{\overset{Cl}{\underset{|}{}}}\overset{\delta-}{}$$
$$I.CH_2.\overset{|}{CH^+}.\overset{\delta+}{CH_2}$$

(**8.10**)

and the two reactions would then give the same ratio of products. Instead, it is found that some migration of iodine to the central carbon atom occurs, (8.7), but that formation of an iodine-bridged intermediate

$$CH_2:CH.CH_2I \xrightarrow{+ClOH}$$

$$Cl.CH_2.CH(I).CH_2OH \text{ and other chloroiodohydrins} \quad (8.7)$$

is by no means complete. The proportion of rearranged product varies with the ionising power of the solvent, amounting to 48 per cent of the iodochloropropanols formed in water, but to only 18 per cent in 70 per cent dioxan. The incompleteness of this rearrangement, and the fact that the two additions give different products, confirms that the intermediates involved are different and probably react with nucleophiles very rapidly, before they have the opportunity of reaching their thermodynamically most stable form.

Additions with accompanying carbocationic rearrangement have been noted in a number of other reactions initiated by electrophilic iodine, as for example in the case of 9,10-dihydro-9,10-ethenoanthracene: (8.8) (Tanner and Brownlee, 1966). Allylic rearrangements, as in the predominantly 1,4-addition of iodine to 1,3-butadiene, (8.9), are also known; but this reaction is probably thermodynamically controlled, since 1,2-

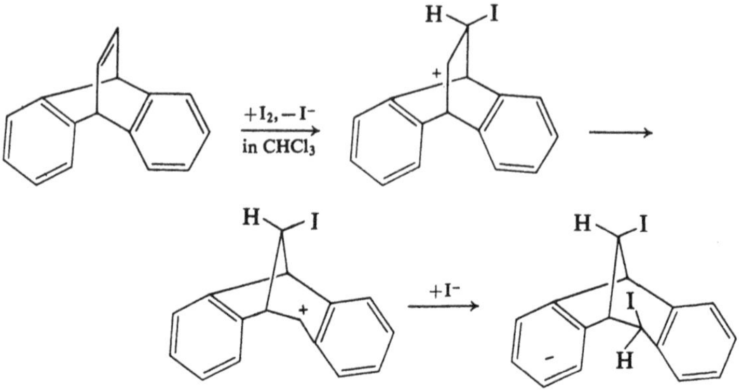

(8.8)

addition seems to prevail in additions in which the product is less labile: (8.10) (Petrov, 1940).

$$CH_2:CH.CH:CH_2 \xrightarrow{+I_2} ICH_2.CH:CH.CH_2I \qquad (8.9)$$

$$CH_2:C(Cl).CH:CH_2 \xrightarrow[-HgI(CH)]{+I_2, +HgO, +MeOH}$$
$$CH_2:C(Cl).CH(OMe).CH_2I \quad (8.10)$$

In an olefinic system strongly activated by conjugative substituents, delocalisation of the positive charge is expected very greatly to reduce bridging, even by iodine, to an adjacent carbocationic centre. Muizebelt (1971) has shown that the isomerisation of *cis*-4,4′-dimethoxystilbene to its *trans*-isomer is catalysed by iodine, and that the products of its reaction with iodine in methanol are not formed stereospecifically, but instead are a mixture of the diastereoisomers expected for reaction through the 'open' carbocation (**8.11**), rotation being allowed about the central carbon–carbon bond before solvent is captured.

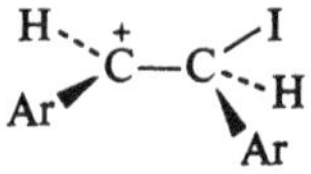

(8.11)

Bridging by iodine to an adjacent carbocationic centre is expected to be of minimal importance in the iodination of aromatic compounds. For reaction with iodine in aqueous solvents, these substrates react at a measurably fast rate only when powerfully activated by conjugative electron-releasing groups, or when the aromatic system itself is a very reactive one. The reactivity sequence PhO⁻ > PhNR₂ > PhOR, and the predominant *para*-substitution, indicate that the rate-determining transition state and any intermediates essential to its formation have much carbocationic character. Berliner (1966) has estimated an approximate ρ^+-value of -13 for molecular iodination, even higher than the corresponding values for the reactions of other molecular halogens.

The kinetics of these iodinations have been studied in considerable detail. The rate equation involves the aromatic compound and one molecule of the iodinating agent. For many aromatic substrates, substantial kinetic isotope effects have been established. This, together with observations of catalysis by nucleophilic bases, indicates that proton loss can become important in the rate-determining step of iodination. As for the corresponding result for the brominations discussed in chapter 7 (§7.3.6; cf. also Berliner, 1964, 1966; Zollinger, 1964) however, this finding is consistent with several possible reaction paths involving as the electrophile either molecular iodine or 'positive iodine' (I^+ or H_2OI^+) in pre-equilibrium with it, and either synchronous or stepwise proton loss. Sequences involving 'positive iodine' were for some time favoured; Grovenstein *et al.* (1962, 1967, 1973) and Shilov *et al.* (Vainshtein, Tomilenko and Shilov, 1963), however, have now established for reactions of azulene, of many but not all aromatic amines, and of a number of typical aryloxide ions that the isotope effect varies with the concentration of iodide ions without any accompanying change in the dependence of rate on the concentration of aromatic compound. Thus the value of k_H/k_D for the iodination of 4-nitrophenol and its 2-deuterio-derivative falls from 5.5 at $I^- = 0.006$ M to 2.3 at $I^- = 0.00001$ M. This result establishes (see also §7.3.6) the two-stage nature of the reaction and the participation of iodide ion in pre-equilibrium with the organic substrate, so that the iodine molecule is implicated as the electrophilic species: (8.11) and (8.12). When the concentration of iodide ions is sufficiently reduced,

$$\text{ArH} + \text{I}_2 \quad \rightleftharpoons \quad [\text{ArHI}]^+ + \text{I}^- \qquad (8.11)$$

$$[\text{ArHI}]^+ \quad \longrightarrow \quad \text{ArI} + \text{H}^+ \qquad (8.12)$$

the rate of the reverse reaction (from right to left) in (8.11) is reduced, the first stage of reaction becomes more nearly rate limiting, and the large primary kinetic isotope effect, expected only for the reaction of (8.12) because only in this stage is a C–H bond broken, becomes diminished.

The mechanism favoured for these iodinations, therefore, is the two-stage process described in (8.11) and (8.12), with the final stage rate-determining in many but not in all cases. In the corresponding iodinations by iodine chloride, the same general considerations apply. The iodination of anisole with iodine chloride in water (Berliner, 1960) follows the expected kinetic form,

$$-\text{d}[\text{ArH}]/\text{d}t = k[\text{ArH}][\text{ICl}]$$

and has a primary deuterium isotope effect, $k_H/k_D = 3.8$, not changed at the lowest accessible concentrations of chloride ions. Again, several mechanisms are consistent with these results, but the one that involves attack by the iodine chloride molecule on the aromatic compound to form an intermediate which then loses a proton in a partly rate-determining step receives support from the kinetics of protodeiodination of iodo-2,4,6-trimethoxybenzoic acid. Here the form of dependence of rate on concentration of chloride ion, which acts as a powerful catalyst for the deiodination, (Batts and Gold, 1964) indicates that the transition state for protodeiodination can have the composition [ArH, ICl]. In accordance with the principle of microscopic reversibility, a transition state of the same composition is available also for iodination by iodine chloride.

8.3 Iodine carboxylates

The kinetic intervention of iodine acetate in aromatic substitutions has been noted in a number of instances, as for example by Grovenstein *et al.* (1973) in the iodination of phenol, and by Chen, Keefer and Andrews (1967) in the iodination of pentamethylbenzene with iodine and mercuric acetate. It has also been considered to be the electrophile concerned in the iodination of benzene in acetic acid catalysed by peracetic acid. (Ogata *et al.*, 1964, 1972). The reaction was found not to be catalysed by mineral acids; and the kinetic form, (8.13), with the rate independent of the concentration of aromatic compound, showed that the rate-limiting step

$$-d[I_2]/dt = k[MeCO_3H][I_2] \qquad (8.13)$$

was the formation of a new iodinating species. A notable feature is the rapidity with which the new reagent attacks benzene, which is usually very sluggish in iodination; it will be remembered that both chlorine acetate and bromine acetate are very powerful electrophilic halogenating agents.

Iodine carboxylates may also be concerned in the acid-catalysed rearrangements of *N*-iodoanilides, (8.14), in aprotic solvents (Bell and Brown, 1936).

$$Ph.N(I)CHO \xrightarrow{\text{HOAc as catalyst}} I.C_6H_4.NH.CHO \qquad (8.14)$$

Bulk concentrations of iodine carboxylates can be prepared by treating a solution of iodine with a suitable silver salt: (8.15). Various solvents can be used; and, though there seem to be few kinetic studies of their reactions

$$AgOAc + I_2 \longrightarrow AgI(\downarrow) + I.OAc \qquad (8.15)$$

with olefinic compounds, these reagents are preparatively of great importance, since they allow the stereospecific functionalisation of both olefinic carbon atoms (for a review see Wilson, 1957). The iodinating agent is often formed *in situ*, and addition proceeds with *anti*-stereospecificity to give the iodocarboxylate adduct, which itself can be isolated if required. Its subsequent reactions can often be controlled so that the functional groups finally incorporated have either *syn*- or *anti*-relationship. In a typical example of Prévost's procedure (cf. Prévost and Lutz, 1934) the olefinic compound is treated with iodine (one molecular proportion) and silver benzoate (two molecular proportions) in a non-polar solvent. Stereospecific *anti*-addition of iodine benzoate is followed by displacement of iodine by the benzoate ion with retention of configuration, the configuration at the attacked carbon atom being maintained by interaction with the neighbouring benzoyloxy-group (structure (**8.12**), fig. 8.1). The overall result is stereospecific *anti*-dibenzoyloxylation of the double bonds.

In Woodward and Brutcher's procedure (1958), iodine and the silver carboxylate (typically silver acetate) are used in equimolecular proportions in acetic acid containing a little water. Under these conditions, the same course is followed initially; but the intermediate (**8.12**) undergoes solvolysis with C–O bond fission to give the hydroxy-ester with overall *syn*-hydroxyacetoxylation of the double bond.

Details of the mechanism of the initial addition are not known with certainty. The intermediates appear to have carbonium ionic character;

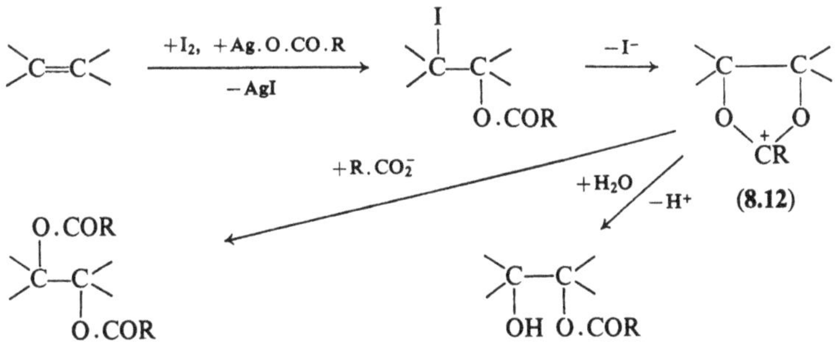

(**8.13**)

Typical 'Prévost' procedure obtained by using I_2 + 2R.CO_2Ag in an aprotic solvent. Overall result is *anti*-addition.

(**8.14**)

Typical 'Woodward' procedure, obtained by using I_2 + R.CO_2Ag in a moist carboxylic acid as solvent. Overall result is *syn*-addition.

Fig. 8.1. Prévost's and Woodward's procedures for hydroxylation of double bonds.

and, as with the additions initiated by molecular iodine, the products can become determined at various stages along the reaction path, depending on the substrate and the conditions of reaction. By using iodine benzoate with butadiene in benzene, for example, Prévost and Lutz (1934; cf. Wilson, 1957) obtained but-1-en-3,4-diol (80 per cent) by 1,2-addition, together with only 4 per cent of but-2-en-1,4-diol by 1,4-addition. Kinetic control of the product ratio, with some addition with rearrangement, is apparent here. With other substrates, products of substitution with allylic rearrangement can be formed: (8.16) (Cambie

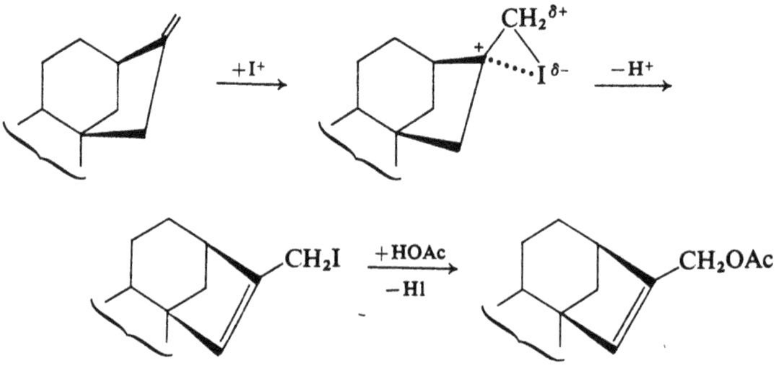

et al., 1974). The formation of small amounts of the products of *syn*-iodocarboxylation of D-glucal triacetate accompanying the *anti*-adducts which are the major products (see later, (8.23); Hall and Manville, 1969) suggests that in this activated olefinic system the solvent may become captured at a late stage, possibly from an iodonium intermediate.

8.4 Reagents having iodine–nitrogen bonds

The characteristics of addition of iodine isocyanate (INCO) to olefinic compounds resemble closely those of the additions of iodine carboxy-lates. The reagent is usually provided by using the unsaturated compound, iodine and silver isocyanate in anhydrous ether. Stereospecific *anti*-addition is found (Hassner *et al.*, 1970); the orientation, though predominantly Markownikoff in direction, is regioselective rather than regiospecific; thus hex-1-ene gives only 70 per cent of $C_4H_9.CH(NCO).CH_2I$. A number of bicyclic compounds, including 9,10-ethenoanthracene (see (8.8)) give products of rearrangement.

Another reagent which gives Markownikoff-oriented *anti*-addition is iodine nitrate, prepared from iodine chloride and silver nitrate, and used in a mixture of pyridine and chloroform (Diner, Worsley and Lown, 1971; Lown and Joshua, 1973). The electrophilic iodinating species here may be the cation, $[I, py_2]^+$. The products are the expected adducts, iodonitrates and iodoalkyl pyridinium nitrates, together with their products of further reaction with the solvent.

Kinetic evidence for the intervention of iodinating species involving I–N bonds comes from the work of Vainshtein, Tomilenko and Shilov (1963). It was shown that in the iodination of aromatic amines, even when no primary kinetic isotope effect was found, the kinetics included a term of the form of (8.17). The involvement of a second molecule of

$$-d[I_2]/dt = k[I_2][ArNR_2]^2 \qquad (8.17)$$

amine suggested the incursion of a mechanism involving a new iodinating agent, possibly of the type $Ar . NR_2I^+$. Schutte and Havinga (1970) have reached similar conclusions.

Electrolytic iodinations of hydrocarbons in acetonitrile have been reported, and may involve the iodinating species $Me . C^+ : NI$ or $Me . CO . NHI$ (Miller, Kujawa and Campbell, 1970).

8.5 Kinetic participation of the nucleophile by attack on the carbocationic centre; mechanisms of the 'Ad3' type

8.5.1 Iodination of acetylenic compounds.
It was seen in earlier chapters that the nucleophilic component of an addition could become involved in the reaction path in a number of ways. One of these involves completion of addition by capture of an intermediate by the nucleophile in the rate-determining stage of the reaction. In principle, the captured intermediate could be any one of the various isomeric forms of molecular complexes, of σ-complexes or of carbocations; and this capture could in favourable cases merge and become synchronous, or nearly so, with electrophilic attack. Such mechanisms can be described as Ad3, and it has already been noted (§7.3.3) that study of the effect of structure on reactivity can reveal a spectrum of transition states in which the forming bond may be dominantly that to the electrophile or to the nucleophile.

So far in this chapter, this mechanism has been neglected, since definite evidence for its incursion has been absent; though it is possible that some of the additions accompanied by solvent capture would better be described in this way. There are several types of iodination, however, in which termolecular reactions seem to be incontrovertibly important. One of these involves the addition of iodine to acetylenes. Miller and Noyes (1952) studied the reverse of this type of reaction, the iodide-catalysed elimination from *trans*-1,2-di-iodoethylene: (8.18). From the

$$\text{\textit{trans}-I.CH:CH.I} + \text{I}^- \rightleftharpoons \text{CH:CH} + \text{I}_2 + \text{I}^- \qquad (8.18)$$

kinetic form and comparison with the *cis*-isomer, it was concluded that the reverse addition must be termolecular and stereospecific; by the principle of microscopic reversibility, forward and reverse reactions carried out under the same conditions must involve the same transition state.

This conclusion has been confirmed by later work. Berliner *et al.* (Wilson and Berliner, 1971; Mauger and Berliner, 1972; Cunningham and Berliner, 1974) have made detailed studies of the iodination of the phenylpropiolate anion (Ph.C:C.CO$_2^-$) and of propiolic acid (HC:C.CO$_2$H) and its anion in water. Two principal reaction paths were identified. The first, of kinetic form

$$-\text{d}[\text{I}_2]/\text{d}t = k[\text{Unsaturated compound}][\text{I}_2]$$

was considered to involve attack by molecular iodine on the acetylenic compound to give a carbocationic intermediate, shown in unbridged form in structure (**8.15**) (fig. 8.2), which then gives the *trans*-di-iodide (**8.16**) together with a derivative of pyruvic acid by competing capture of a solvent residue. The second, dominant at high concentration of iodide ions, probably involves a transition state such as is shown in structure (**8.17**), which may be reached in either synchronous or stepwise fashion. It gives stereospecifically only the *trans*-di-iodide. The effect of change in structure on the rate of reaction through this path suggested that a

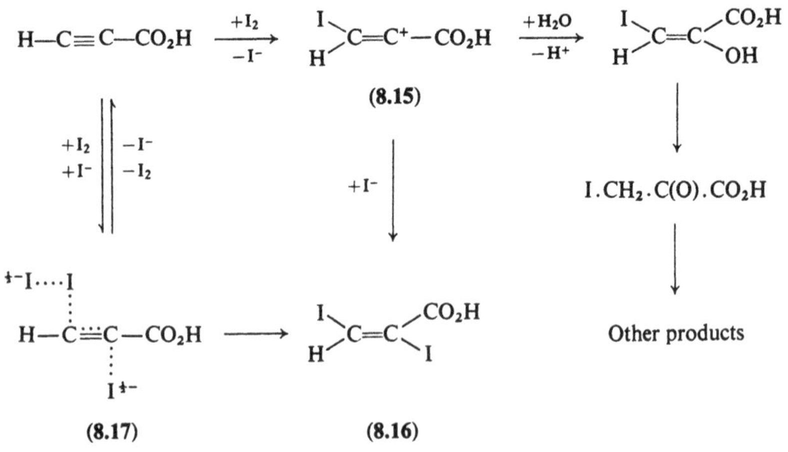

Fig. 8.2. Probable reaction paths in the iodination of propiolic acid and related compounds with iodine in water at 25 °C.

spectrum of transition states may be involved, with neither the electrophilic nor the nucleophilic component unambiguously dominant.

8.5.2 Intramolecular kinetic participation by nucleophilic neighbouring groups. Mechanistically analogous to the kinetic participation by iodide ion in the addition of iodine to acetylenic compounds is the intramolecular participation by neighbouring hydroxyl and carboxylate ion groups in the reactions of iodine with unsaturated alcohols and acids. The relative rate coefficients for the iodination of a series of alcohols $CH_2:CH$. $(CH_2)_nOH$ by iodine in aqueous potassium iodide were found by Williams, Bienvenüe-Goetz and Dubois (1969) to be 1, 2.1, 200 and 34.5 for $n = 1, 2, 3$ and 4 respectively. The main product of the reaction of the pentenol ($n = 3$) was shown to be the derivative of furan shown in (8.19). If the ring closure occurred after the rate-determining stage of the

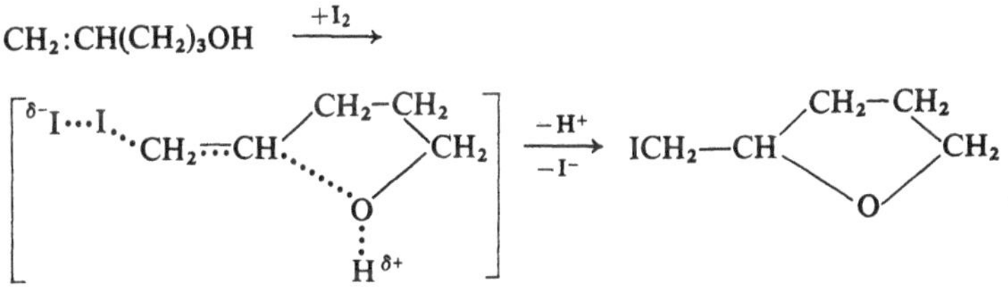

$$CH_2:CH(CH_2)_3OH \xrightarrow{+I_2}$$

(8.18)

reaction, the rate of iodination would be expected to increase from $n = 1$ to $n = 4$. The observed maximum at $n = 3$, therefore, indicates that for this compound the hydroxyl group is participating in the rate-determining stage, probably through a transition state such as is shown in structure (**8.18**).

The corresponding influence of the carboxylate ion group is shown in iodolactonisation of the salts of unsaturated acids. This is synthetically an important process, in which iodine and oxygen functions are introduced with *anti*-stereospecificity and can often be transformed into new functional groups. A simple example is given in (8.20). Staninets

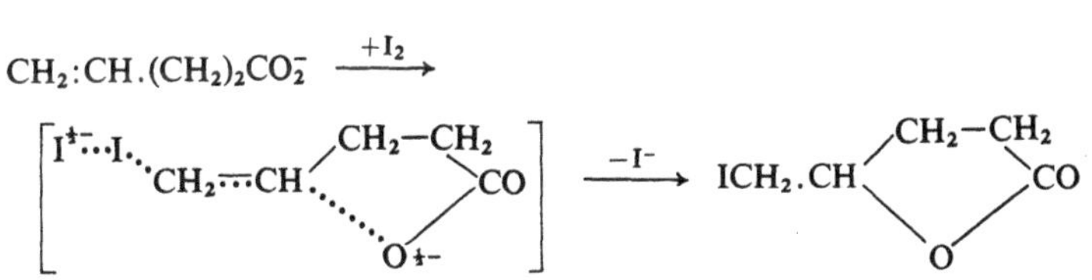

$$CH_2:CH.(CH_2)_2CO_2^- \xrightarrow{+I_2}$$

(8.19)

TABLE 8.1 *Rates of iodolactonisation of unsaturated carboxylate anions with iodine in aqueous potassium iodide at 20 °C*

Substrate	$k_2/l \text{ mol}^{-1} \text{ s}^{-1}$	Substrate	$k_2/l \text{ mol}^{-1} \text{ s}^{-1}$
$CH_2:CH.CO_2^-$	0	$HC\vdots C.CH_2.CH(CO_2^-)_2$	3.6
$CH_2:CH.CH_2.CO_2^-$	0.01	$CH_2:CH.CH_2.CH(CO_2^-)_2$	169
$CH_2:CH.(CH_2)_2 \ CO_2^-$	84	$Me.CH:CH.(CH_2)_2.CO_2^-$	142
$CH_2:CH.(CH_2)_3CO_2^-$	11	$Ph.CH:CH.(CH_2)_2CO_2^-$	61
$CH_2:CH.(CH_2)_4CO_2^-$	0.17	$Cl.CH:CH.(CH_2)_3.CO_2^-$	0.025
$CH_2:CH.(CH_2)_5CO_2^-$	0		

and Shilov (1971) have reviewed the reaction and have discussed kinetic evidence relating to the mechanism. The simple kinetic form

$$-d[I_2]/dt = k[\text{Unsaturated compound}][I_2]$$

is observed, and the rate coefficients for some typical cyclisations are given in table 8.1. Ring closure normally proceeds most easily to form a five-membered ring, as would be expected from the rate sequence given in the table; but competition can occur with the formation of rings of other sizes, and both steric and polar effects are concerned in determining the orientation of the product. The reactions are kinetically controlled, and probably involve cyclic transition states such as (**8.19**). The similar reaction of pent-4-enoic acid ($CH_2:CH.CH_2.CH_2.CO_2H$) in chloroform is of the second order in iodine (Amaral and Melo, 1973).

When iodinated in a two-phase system of ether and water, the salts of $\beta\gamma$-unsaturated acids can give β-lactones under conditions of kinetic control: (8.21) (Barnett and Sohn, 1972). It is not quite clear whether

$$CH_2:CH.CMe_2.CO_2^- \xrightarrow[-I^-]{+I_2} I.CH_2.CH\overset{CMe_2}{\underset{O}{\diagdown\diagup}}CO \qquad (8.21)$$

the synchronous mechanism of iodolactonisation is operative here also, since salts of these acids tend to be rather unreactive (see table 8.1). Another convenient procedure for iodolactonisation involves the addition of iodine to an unsaturated thallium(I) carboxylate in ether at 20 °C (Cambie *et al.*, 1974).

8.5.3 Iodine fluoride and iodine azide as reagents for addition. The additions of iodine fluoride and of iodine azide have much in common with those of other molecular iodinating agents, but there are certain differences of behaviour which suggest the incursion of termolecular mechanisms. The iodofluorination of olefinic compounds can be carried out by using potassium fluoride and iodine in acetonitrile (Krespan, 1962), or N-iodosuccinimide and potassium fluoride in a mixture of methylene chloride and tetrahydrofuran (Bowers, Denot and Becerra, 1960). From cholesterol, the 6β-iodo-5α-fluoride was obtained, (8.22),

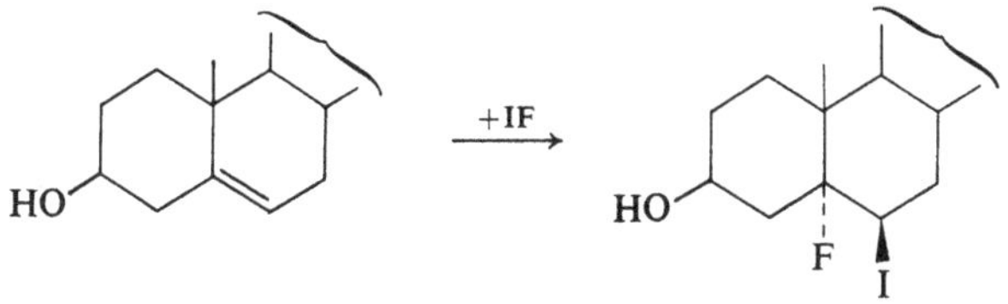

$$\xrightarrow{+\,\mathrm{IF}} \qquad (8.22)$$

by *anti*-addition with Markownikoff orientation. This result is unexpected only because bromine fluoride under similar conditions gave instead the 5α-bromo-6β-fluoride, with electrophilic attack initiated on the opposite face of the molecule. It was suggested that an equilibrium is set up between α- and β-iodonium intermediates, the product composition then being determined at rather a late stage on the reaction path. An alternative possibility is that a mechanism of the termolecular type is under observation, with rate-determining attack by fluoride ion on a complex between the iodinating agent and the β-face of the cyclic olefin.

Hall and Manville (1969) have recorded similar procedures in which iodine and silver fluoride in benzene, or N-iodosuccinimide and anhydrous liquid hydrogen fluoride, are used for additions to such unsaturated derivatives of carbohydrates as D-glucal triacetate, (**8.20**). The products

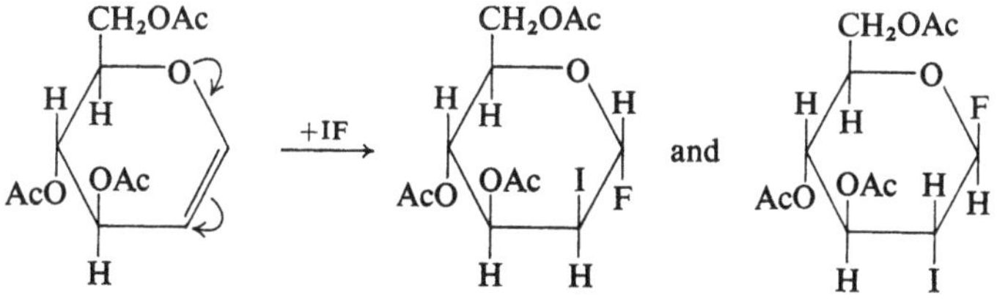

$$\xrightarrow{+\,\mathrm{IF}} \qquad \text{and} \qquad (8.23)$$

(**8.20**)

again are nearly exclusively those of *anti*-addition, here to either face of the molecule, with orientation determined in the expected way by electron release from the unsaturated substituent (see arrows in structure (**8.20**)). It will be recalled that the corresponding additions of iodine carboxylates were not quite so stereospecific.

In a similar way, iodine azide can be supplied for iodination of olefinic compounds by using iodine and sodium or other metallic azides in an aprotic solvent. The resulting β-iodoalkyl azides serve as useful starting materials for syntheses involving vinyl azides, amines, aminolactones, aziridines and other nitrogen-containing heterocycles. These reactions, which are remarkably regiospecific and stereospecific, have been reviewed by Hassner (1971). An example is given in (8.24). The resulting Markownikoff orientation and *anti*-stereochemistry suggests either that the

$$
\underset{Ph}{\overset{H}{\diagdown}} C{=}C \overset{H}{\underset{D}{\diagup}} \quad \xrightarrow{+IN_3} \quad \underset{Ph}{\overset{H}{\diagdown}} \overset{N_3}{\underset{I}{\overset{|}{C}}}{-}\overset{}{\underset{}{C}} \overset{H}{\underset{D}{\diagup}} \tag{8.24}
$$

electrophilic iodine bridges very effectively to the carbocationic centre in an unsymmetrical intermediate (e.g. as shown in structure (**8.21**)), or that a termolecular mechanism of reaction between the olefinic substrate, iodine or iodine azide, and azide ion is under observation.

The second of these hypotheses is supported by the fact that products of incorporation of solvent appear less prominently in reactions involving iodine azide than in those involving other molecular iodinating agents. Kinetic investigation of the role of the nucleophile in these additions would be valuable; the possible incursion of mechanisms of the Ad3 type has been considered also by Hayward (1973), who has used arguments similar to those presented above.

$$
\underset{Ph}{\overset{H}{\diagdown}} \overset{+}{\underset{\diagdown}{C}}{-}\overset{\delta+}{\underset{I\,\delta-}{\overset{|}{C}}} \overset{H}{\underset{D}{\diagup}}
$$

(**8.21**)

When steric hindrance to attack by the nucleophile is sufficiently great, even iodine azide gives addition with carbocationic rearrangement, as is shown for the more general case (addition of I.Nu) in (8.25).

$$
Ph_3C.CH{:}CH_2 \quad \xrightarrow{+I.Nu} \quad Ph_2C(Nu).CH(Ph).CH_2I \tag{8.25}
$$

8.6 Iodination in strongly acidic conditions

For the iodination of relatively unreactive substrates, a number of preparative methods of iodination have been developed which are effective only in highly acidic conditions. Iodine is often used together with a reagent which ensures a favourable position of equilibrium. Thus Derbyshire and Waters' procedure (1950) involves treatment of the aromatic compound with iodine, silver sulphate and sulphuric acid. The effective reagent may be I^+, or more probably a co-ordinated form of this (perhaps ISO_3H^+).

Other procedures involve the provision of an oxidising agent, and this often has the dual function of removing hydrogen iodide and providing a new active iodinating species. Iodination by iodine in mixtures of nitric acid and acetic acid has been investigated mechanistically by Butler and Sanderson (1974). The dependence of the rate on the concentration of acid and of dinitrogen tetroxide (N_2O_4) was interpreted as indicating that the electrophile is the $[INO_2H]^+$ cation, formed by protonation of nitrogen dioxide to give NO_2H^+, followed by reaction of this with iodine.

When iodine and an iodate in relative molecular proportions 7:1 are dissolved in concentrated sulphuric acid, a deep brown solution containing the I_3^+ cation is formed: (8.26). The resulting solution is a powerful

$$7I_2 + IO_3^- + 9H^+ = 5I_3^+ + 3H_3O^+ \tag{8.26}$$

reagent for iodination, often giving polysubstituted products. Arotsky *et al.* (1968, 1973) have shown that the response of rate to change in structure over a range of deactivated aromatic compounds corresponds to a value of $\rho^+ = -6.4$, similar to that found for bromination by 'positive bromine' and for nitration by the nitronium ion. The kinetic isotope effect was found to be $k_H/k_D = 2.1$ for benzoic acid and 3.4 for nitrobenzene, so proton loss can participate in determining the rate of this iodination. It is not certain whether the I_3^+ cation acts directly on the aromatic nucleus, or whether it releases I^+ or some other reactive species in pre-equilibrium before reaction.

Suzuki (1971) has recommended the use of periodic acid and iodine in acetic acid with sulphuric acid as catalyst for some preparative iodinations. 2,4,6-Trimethylbenzoic acid gave the 3-iodo-derivative; pentamethylbenzoic acid gave iodopentamethylbenzene by iododecarboxylation: (8.27).

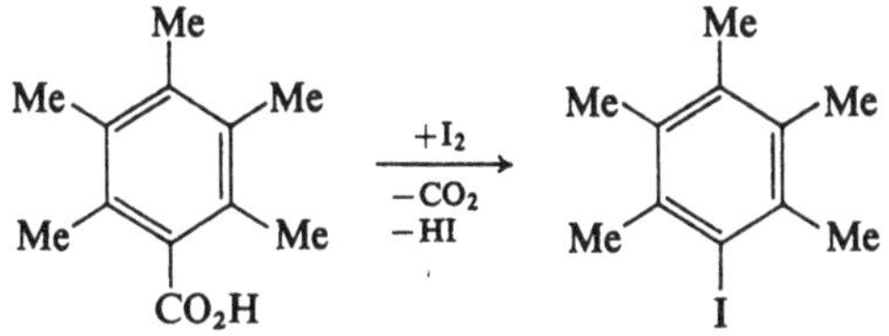

8.7   Iodination in the *ortho*-position

It is general preparative experience that iodination *ortho* to bulky substituents is even more difficult than bromination, as is consistent with the greater bulk of iodine. Direct mechanistic comparison is difficult, however, since the rate-determining stages for bromination and iodination usually differ, in that only in the latter does the stage of proton loss become rate determining. The iodination of phenol normally gives *p*-iodophenol, but conditions have been reported by which the *ortho*-isomer can be obtained in good yield: iodine is treated with sodium phenoxide in anhydrous xylene (Hunter and Budrow, 1933). Possibly this involves the formation of Ph.OI and its intramolecular rearrangement.

9 Miscellaneous electrophilic halogenations, particularly of compounds containing 'active hydrogen'

'Change is not made without inconvenience...' (R. Hooker)

9.1 Introduction

Electrophilic replacement by halogen at saturated carbon is usually difficult, though in certain circumstances special structural features make it possible. One of these is internal ring strain. Thus although many ring openings of cyclopropanes are homolytic in character, heterolytic reactions can also occur (Turnbull and Wallis, 1956; Gordon, 1967). Cyclopropane and its derivatives react with bromine in acetic acid with a kinetic form which resembles that for other heterolytic brominations, and structural effects suggest that bromine acts as an electrophile. The products have not been studied extensively; rearrangements can accompany these reactions under some circumstances, but probably the main reaction gives 1,3-dibromocyclopropanes. Dusseau, Schaafsma and De Boer (1970) have shown, for example, that 1-phenylcyclopropanone dimethylacetal (9.1) with bromine in sulphur dioxide at −60°C gives the carbocation (9.2) stabilised by conjugation with the methoxy-groups. This then decomposes as is shown in the sequence (9.1).

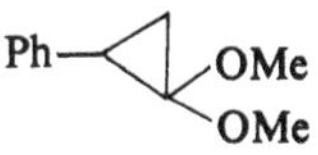

$$Ph-\underset{\textbf{(9.1)}}{\triangleleft}\overset{OMe}{\underset{OMe}{}} \quad \xrightarrow[-Br^-]{+Br_2}$$

$$PhCH(Br).CH_2.C^+(OMe)_2 \underset{\textbf{(9.2)}}{\quad} \xrightarrow[-MeBr]{+Br^-} PhCH(Br)CH_2CO_2Me \quad (9.1)$$

Electrophilic halogenations become possible also if the leaving group is one which can be displaced sufficiently easily. A general review of halogenodemetallations has been given by Abraham (1973). In the displacement of chromium by electrophilic bromine shown in (9.2) an

177

open, rather than a cyclic, transition state seems to be involved, since free bromide ion rather than the stable complex ion $[(H_2O)_5CrBr]^{2+}$ is formed (Espenson and Williams, 1974). Various exchanges involving

$$[R.Cr(H_2O)_5]^{2+} \xrightarrow[\text{in aq. HClO}_4]{+Br_2, +H_2O} R.Br + [Cr(H_2O)_6]^{3+} + Br^- \qquad (9.2)$$

electrophilic iodine, on the other hand, may involve cyclic transition states (Noyes and Koros, 1971), in which both electrophilic and nucleophilic attack on iodine can be envisaged.

Another way in which reaction with electrophilic halogen may effect replacement at saturated carbon is illustrated by the chlorinolysis of alkyl aryl sulphides and sulphenate esters: (9.3) and (9.4). Tetraligant

$$Ph.CH(Et).SPh \xrightarrow[\text{in HOAc}]{\text{with Cl}_2} \begin{cases} Ph.CH(Et)Cl \\ Ph.CH(Et)OAc \end{cases} \text{ and other products} \qquad (9.3)$$

$$Ph.CH(Et).O.SAr \xrightarrow{+Cl_2} Ph.CH(Et)Cl + Ar.SO.Cl \qquad (9.4)$$

sulphides, $Ph.CH(Et).SCl_2Ph$ and $Ph.CH(Et).OSCl_2Ar$, are believed to be intermediates which then allow nucleophilic inter- or intra-molecular replacement at saturated carbon in completion of the reaction. Related sequences may be concerned in the halogenation of sulphoxides (Iriuchijima and Tsuchihashi, 1973). It is obviously difficult to classify mechanistically these complex sequences (cf. de la Mare and Swedlund, 1973), which start with an electrophilic attack at an atom other than carbon or hydrogen.

By use of a sufficiently strong base, of course, any hydrogen can in principle be removed from carbon, and the resulting carbanion can then react with an electrophilic halogenating agent. This reagent must be stable enough to survive the basic conditions long enough to react with the carbanion; it will be seen that, depending on the circumstances, hypohalite ions, organic hypohalites, *N*-halogeno-compounds and the halogens themselves can all function in this way. Examples can be drawn from the chemistry of substitution both at saturated and at unsaturated carbon. Thus Mach and Bunnett (1974) have shown that the reaction of 1,3,5-tribromobenzene with t-butyl hypobromite in *N,N*-dimethyl-formamide requires a basic catalyst; the sequence shown in (9.5) is followed. Again, cyclopentadiene and similar compounds are chlorinated by the hypochlorite ion in alkaline solution to give perchlorocyclopentadiene, the first stage being that shown in (9.6) (Straus, Kollek and Heyn, 1930).

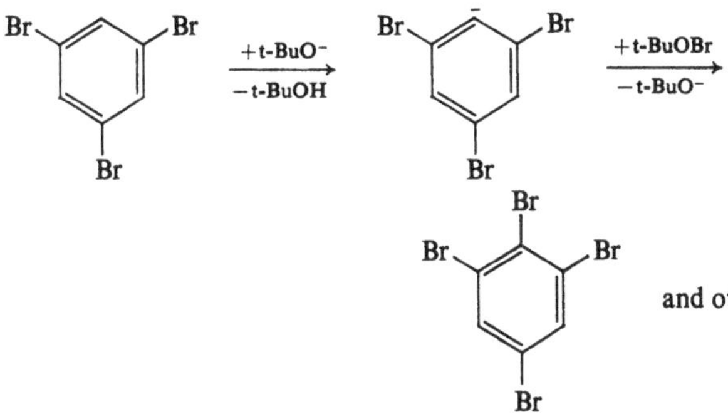

and other products (9.5)

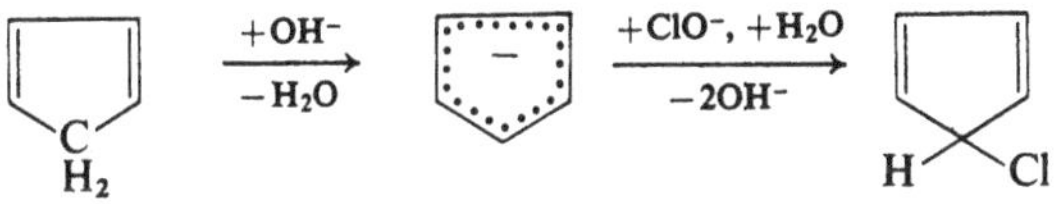

(9.6)

Base-catalysed halogenations of sulphones by molecular halogens in water are similar; the mechanisms have been studied by Bell and Cox (1971), who have reviewed the earlier literature. For monosulphones, the rates are independent of the halogenating species; so the sequence is that of (9.7), with the rate of formation of the carbanion (v_1) rate determining.

$$R_2CH.SO_2.R' \underset{+BH^+, -B \ (v_{-1})}{\overset{+B, -BH^+ \ (v_1)}{\rightleftharpoons}} \underset{(9.3)}{R_2C^-.SO_2.R'} \underset{(v_2)}{\overset{+X_2, -X^-}{\longrightarrow}} R_2C(X).SO_2.R'$$

(9.7)

No catalysis by acid was detectable, and by using disulphones (e.g. $(EtSO_2)_2CHMe$) at a suitable acidity, the rates of the competing reactions of the carbanion (9.3) to give starting material and product (v_{-1} and v_2 respectively) could be adjusted so that the rate became dependent on the halogen. An estimate could then be made that the anion reacts with molecular halogen nearly at the diffusion rate.

Kinetic situations in which the removal of 'active hydrogen' by a base determines the rate of a subsequent reaction have been known since Lapworth (1904) showed that the rate of halogenation of acetone is independent of the nature or concentration of the halogen, and is subject to catalysis both by acids and by bases. The former type of catalysis normally follows the sequence of (9.8), whereas the latter follows that

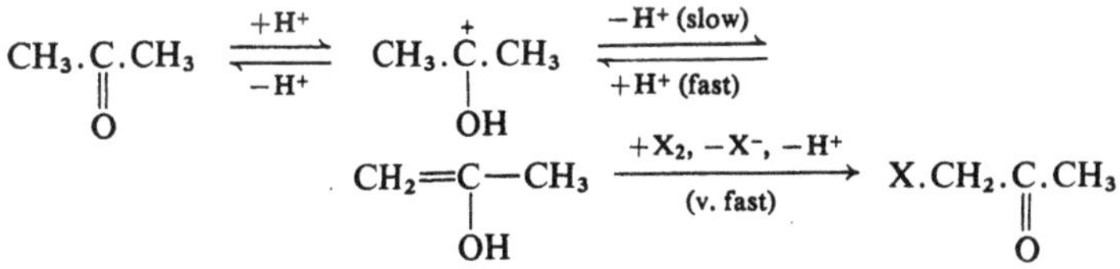

(9.8)

of (9.9). Under catalysis by acid, the intermediate attacked by the halogenating agent is the very reactive olefinic compound, the enol (9.4); it gives the product of substitution with rearrangement (the S_E2'

$$
\begin{array}{c}
\text{CH}_3\text{.C.CH}_3 \\
\parallel \\
\text{O}
\end{array}
\quad
\underset{\substack{+\text{BH}^+, -\text{B} \\ \text{(fairly fast)}}}{\overset{\substack{+\text{B}, -\text{BH}^+ \\ \text{(slow)}}}{\rightleftharpoons}}
$$

$$
\left[
\begin{array}{c}
{}^-\text{CH}_2\!-\!\underset{\substack{\parallel \\ \text{O}}}{\text{C}}\!-\!\text{CH}_3 \;\longleftrightarrow\; \text{CH}_2\!=\!\underset{\substack{\mid \\ \text{O}^-}}{\text{C}}\!-\!\text{CH}_3
\end{array}
\right]
\xrightarrow[\text{(v. fast)}]{+\text{X}_2, -\text{X}^-}
\text{X.CH}_2\text{.}\underset{\substack{\parallel \\ \text{O}}}{\text{C}}\text{.CH}_3
\quad (9.9)
$$

$$(9.5)$$

reaction). With catalysis by base, on the other hand, the even more reactive mesomeric anion (9.5) is formed. In the sections following, acid-catalysed halogenations are treated first, in rather more detail than their base-catalysed analogues.

9.2 Kinetic forms for the acid-catalysed halogenation of ketones in hydroxylic solvents

Compounds having a hydrogen atom α to a carbonyl group can ordinarily undergo prototropic rearrangement to the enol tautomer relatively rapidly under catalysis by acids. Even simple ketones such as acetone contain a little of the enol form in the equilibrium mixture; the presence of nearby electron-withdrawing substituents (particularly those which can conjugate with the new double bond, as in β-diketones and keto-esters) tends to shift the equilibrium in favour of the enol form, and to make the rate of attainment of equilibrium more rapid. The spontaneous tautomeric change of keto to enol form may be relatively slow, so that pure samples of the thermodynamically unstable tautomer can sometimes be prepared; but to do this is often difficult, because of autocatalysis of the rearrangement and the influence of traces of adventitious catalysts.

Among the consequences of enolisation is the possibility that the enol form will react with a source of electrophilic halogen. Some of the preparative aspects of these processes, in which the halogen molecules, interhalogen compounds, perchloryl fluoride ($FClO_3$), sulphuryl chloride, copper(II) chloride and bromide, and N-halogenoamides are among the useful electrophiles, have been reviewed by House (1965). Mechanistic aspects are summarised by Bell (1973), by Ingold (1969), and in many textbooks of physical organic chemistry. The most commonly observed kinetic form for reaction involving molecular halogen is that of (9.10). The fact that the rate of reaction is of the zeroth kinetic

$$-d\,[\text{Halogen}]/dt \;=\; k[\text{Ketone}][\text{H}^+] \tag{9.10}$$

order in halogen (i.e. is independent of the concentration and of the nature of the halogen) is consistent with the view that the slow rate-determining stage of the reaction is the formation of the enol, which then reacts with halogen more rapidly than it reverts to ketone: (9.8). This view is supported by the fact that in a variety of representative cases it has been shown that the rate of incorporation of deuterium from the solvent at the α-carbon atom, the rate of loss of optical activity if chirality at the α-carbon atom is lost in the enol (as when the α-carbon atom is 'asymmetric') and the rate of bromination are all equal under the same conditions of solvent and catalyst.

The simplest form of catalysis by acids involves the pre-equilibrium proton transfer shown in the first stage of (9.8). That this stage is not rate determining is established through the observation of a reverse solvent deuterium isotope effect, $k_{\text{D}_2\text{O}}/k_{\text{H}_2\text{O}} = 2.0$ for the acid-catalysed bromination of acetone in water. A reverse isotope effect of this kind arises because positively charged deuterio-acids in deuterated solvents are weaker acids than their protio-analogues; so in the present instance the deuterated conjugate acid, $[\text{Me}_2\text{C}:\text{OD}^+]$, is formed in higher concentration than $[\text{Me}_2\text{C}:\text{OH}^+]$, and hence provides enol at a higher rate.

The fact that even at relatively high concentrations of acid the rate follows the stoicheiometric concentration of acid rather than the Hammett acidity function, h_0, was for some time taken as confirming that the loss of a proton from carbon in the rate-determining stage of the reaction is assisted by a water molecule acting as a base. Complications associated with this view have been discussed carefully by Rochester (1970), since deviations from the ideal relationship have been noted in a number of cases. There seems little reason to doubt, however, that the water molecule can function in this way. Support comes from the fact that catalysis by 'general acids' is often observed through the incursion of a kinetic term of the form given in (9.11). The most natural interpretation of this

$$-d[\text{Halogen}]/dt \;=\; k[\text{Ketone}][\text{H}.\text{Nu}] \tag{9.11}$$

kinetic term is that the reaction sequence is now that of (9.12). Carboxylic

$$\text{Ketone} + \text{H}^+ \;\rightleftharpoons\; [\text{Conjugate acid}]^+ \xrightarrow{\;+\,\text{Nu}^-\;} \text{Enol} + \text{H}.\text{Nu} \tag{9.12}$$

acids in water, and hydrogen halides in acetic acid (Cieciuch and Westheimer, 1963), are among the general acids which are known to be effective. The catalytic power in certain cases follows the kinetic rather than the equilibrium nucleophilicity; bromide ion is a better catalyst than chloride ion in slightly aqueous acetic acid, despite the greater dissociation of hydrogen bromide than of hydrogen chloride.

Intramolecular assistance to such enolisations by carboxylic acid groups is known also, and has been reviewed recently by Bell and Page (1973). Comparison of the rates of iodination of 2-oxobicyclo[2,2,2]octane-1-carboxylic acid and its methyl ester, (9.6) with R = H and Me respectively, in various buffered and self-buffered solutions in water shows the incursion of a kinetic term much larger for the acid than for its methyl ester. It was concluded that the transition state for enolisation must be stabilised intramolecularly, possibly as in (9.7); the exact location of the catalysing proton could not be identified unambiguously.

(9.6)   (9.7)

Acid-catalysed halogenation having the kinetic form of (9.10), of zeroth order in halogen, will be observed only if the reaction of the intermediate enol (9.4) with halogen is much faster than the rate of its reversion to starting material. With the concentration of halogen sufficiently low, and the acidity sufficiently high, this condition may not be satisfied, and the kinetic form then changes; so that in the limit the rate becomes dependent on the halogen and independent of acidity ((9.13),

$$-d[\text{Halogen}]/dt = k[\text{Enol}][X_2] = k'K[\text{Ketone}][X_2] \qquad (9.13)$$

K being the equilibrium constant between keto and enol form). Bell and co-workers (Bell, 1973) and Dubois and co-workers (Toullec and Dubois, 1973) have realised this type of kinetic situation in a number of examples. If the equilibrium constant for the keto–enol interconversion can be estimated, the second-order velocity constant for reaction of halogen with the enol can then be evaluated. For very highly activated systems, as with the enol of acetone ($\text{MeC(OH)}:\text{CH}_2$), the rate approximates to that of a reaction occurring at every collision of the enol and halogen, and hence to that of a reaction rate controlled by diffusion. For less reactive enols, however, lower rates are reported, in the region 10^5 dm^3 mol^{-1} s^{-1}. The rates of bromination of the corresponding enol ethers (e.g. of dimedone enol methyl ether: (9.14), structure (**9.8**)) in water are similar, though perhaps slower by a small factor. A rate difference in a similar direction for the bromination of phenol and anisole has been attributed to O–H hyperconjugation, which here allows partial displacement of the proton in the transition state (**9.10**) (de la Mare, 1959; de la Mare and Dusouqui, 1967; see also §7.3.1).

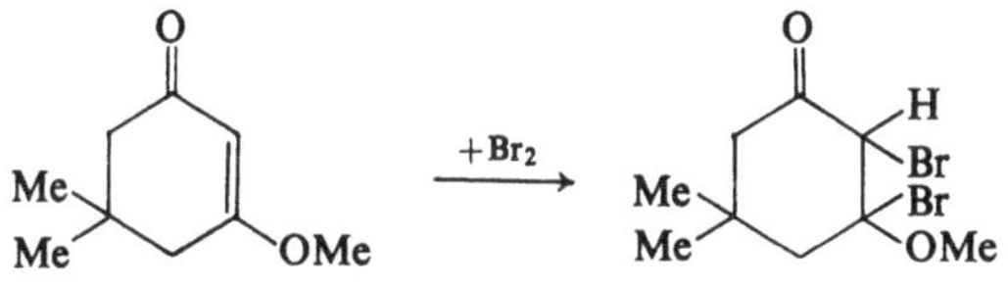

$$(9.14)$$

(9.8)   **(9.9)**

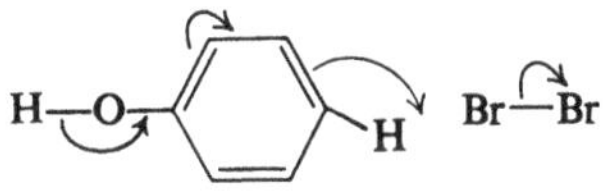

(9.10)

Marshall and Roberts (1971) showed spectrophotometrically that the reaction of dimedone enol methyl ether (**9.8**) with bromine gives first an adduct in which conjugation of the double bond with the oxygen function is destroyed, probably as in (**9.9**). In the analogous bromination of an enol, the loss of a proton from oxygen is expected to be so fast that it is usually treated as being concerted with the attack by bromine. This description is reasonable for reaction in hydroxylic solvents; the situation is not quite so certain when the solvent is aprotic and has limited power of proton acceptance. An example in which a new reaction appears to compete with the normal halogenodeprotonation has been reported by Teo and Warnhoff (1973), and relates to just such a solvent. The chlorination of 4-t-butylcyclohexanone (**9.11**) in carbon tetrachloride at 0 °C (fig. 9.1) was shown to give some 30 per cent of a 2,6-dichloro-4-t-butylcyclohexanone (**9.12**), the formation of which was attributed to the occurrence of the substitutive attack by chlorine on the enol (**9.13**) with rearrangement to give the chloro-enol (**9.14**) by an S_E2' reaction in competition with the normal chlorination to give 2-chloro-4-t-butyl-cyclohexanone (**9.15**), formally also an S_E2' reaction. Whether the stages leading to the dichlorocyclohexanone are synchronous or can be preceded by addition is not known.

Two other special mechanisms for the acid-catalysed halogenation of ketones have been established by kinetic measurements. The first involves the halogenation of acetone in water, catalysed by ammonium ions.

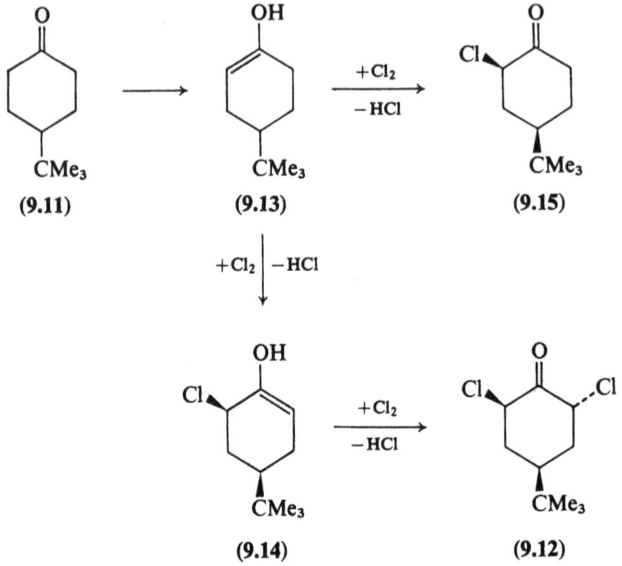

Fig. 9.1. Proposed reaction path in the mono- and di-chlorination of 4-t-butyl-cyclohexanone in carbon tetrachloride at 0 °C.

Bender and Williams (1966) showed that the cations of tertiary amines were not effective; so it was concluded that the catalysis derives from the sequence of (9.15), in which a protonated enamine is an intermediate.

$$\text{Me}_2\text{C:O} \xrightleftharpoons[]{+\text{R.NH}_3^+, -\text{H}_2\text{O}} \text{Me}_2\text{C:NHR}^+ \xrightarrow[\text{slow}]{-\text{H}^+}$$

$$\text{Me.C(:CH}_2).\text{NHR} \xrightarrow[\text{fast}]{+\text{I}_2} \text{products} \quad (9.15)$$

The second derives from Bell and Yates' report (1962) that the chlorination of acetone with chlorine in water includes at high acidities a kinetic term of the form shown in (9.16), with a rate faster than that of the

$$-\text{d[Cl.Nu]}/\text{d}t = k[\text{Acetone}][\text{H}^+][\text{Cl.Nu}] \quad (9.16)$$

formation of the enol. This kinetic form could arise through an attack of protonated hypochlorous acid (or its kinetic equivalent) on the acetone molecule with synchronous displacement of a proton. Further details are not known.

9.3 Structural influences in the acid-catalysed halogenation of ketones

The effects of structure on the rates and products of acid-catalysed halogenation of ketones involve complicated interactions between features related to the polar effects of substituents and to stereochemistry. For acyclic compounds, it is found that electron-releasing substituents somewhat accelerate and electron-withdrawing substituents somewhat retard the rate of bromination. The response of rate to structure is not great; the value of ρ is -0.2 for substituents in the benzyl group of benzyl phenyl ketones, $R.C_6H_4.CH_2.C(:O).Ph$, and -0.6 for substituents in the phenyl group of substituted acetophenones, $R.C_6H_4.C(:O).CH_3$ (Fischer, Packer and Vaughan, 1962). These trends are consistent with the finding that when competing hydrogen atoms are available for substitution, this is disfavoured adjacent to inductively electron-withdrawing groups, and favoured adjacent to hyperconjugatively electron-releasing substituents, which can stabilise the double bond as it develops. Equations (9.17) and (9.18) illustrate this by showing the predominant reactions occurring in the bromination of bromoacetone and of butan-2-one.

$$Br.CH_2.CO.CH_3 \xrightarrow{+Br_2, \, -HBr} Br.CH_2.CO.CH_2Br \qquad (9.17)$$

$$Me.CH_2.CO.CH_3 \xrightarrow{+Br_2, \, -HBr} Me.CH(Br).CO.CH_3 \qquad (9.18)$$

These orientational results refer to conditions in which the proportions of isomeric products reflect the relative rates of enolisation. If the conditions are changed so that the halogen molecule enters the rate equation (9.13), a different orientational result is expected, the effect of electron release becoming more pronounced. Deno and Fishbein (1973) have illustrated this important point by studying the products of chlorination and bromination of butan-2-one in aqueous mineral acids. At low acidity, the ratio of 3-halogeno- to 1-halogeno-butan-2-one is 2.6; it changes to 11 for chlorination, and to 9 for bromination, when the acidity is sufficiently great; under these circumstances the rate of acid-catalysed reversion of the enol to the ketone has been increased sufficiently by increasing the acidity of the medium.

Another way in which orientation of substitution can be changed involves change in the solvent. Gaudry and Marquet (1970; cf. Jasor *et al.*, 1973) have shown that the acid-catalysed bromination of cyclohexyl methyl ketone in carbon tetrachloride gives mainly substitution in the cyclohexyl ring, whereas when the solvent contains methanol the main product is bromomethyl cyclohexyl ketone. Similarly, 2-bromo-cyclohexanone when brominated in ether gives a mixture of 2,2-dibromo- and 2,6-dibromo-cyclohexanone, whereas the latter is the main product in methanol or when the ketal of cyclohexanone is brominated. It seems likely by analogy with the direction of bromination of bromoacetone, (9.17), that the 2,6-dibromide results from kinetic control of the direction of enolisation, whereas the 2,2-dibromide is formed when the product ratio is controlled by the ratio of isomeric enols formed in equilibrium proportions; a somewhat different interpretation is favoured by Garbisch (1965) in his discussion of this and analogous brominations.

The use of other reagents for halogenation can also alter the ratios of isomeric products. By the use of copper(II) chloride in dimethylformamide, Kosower *et al.* (1963) obtained 55–70 per cent of 3-chlorobutan-2-one from butan-2-one (cf. (9.18)). On the other hand, chlorination of the same ketone with sulphuryl chloride gives mainly 3,3-dichlorobutan-2-one (Wyman and Kaufman, 1964). Both these reactions are thought to involve enolisation followed by electrophilic chlorination, but further details are speculative.

Cyclic systems often provide geometric constraints on systems involving double bonds, and those on a substituted ketone, (**9.16**), are different from those on the derived enol, (**9.17**). In the ketone, free rotation about the central C–C bond is permitted; in the enol, it is restricted, and the groups attached to the double bond must be coplanar if strain is to be avoided. The most extreme consequence of this is that acid-catalysed formation of the enol, and hence bromination of the ketone, may be almost completely excluded at a particular position. This is the situation in camphorquinone (**9.18**) where substitution does not occur because of the strain which would be introduced by the formation of a double bond at the bridgehead in contravention of Bredt's rule (Fawcett, 1950).

(9.16)   (9.17)   (9.18)

Orientational and stereochemical aspects of acid-catalysed brominations have been studied extensively in relation to the halogenation of steroidal ketones. Study of these brominations can be complicated by the intervention of acid-catalysed rearrangements which can result in geometrical or positional isomerisation of the products. It is sometimes difficult, therefore, to know whether or not the material isolated was formed under kinetic control; comparison with the reactions of the corresponding enol ethers or enol esters can be helpful diagnostically, since these compounds need milder conditions and do not require catalysts for the halogenations.

The reactions have been reviewed by Kirk and Hartshorn (1968) and by Fieser and Fieser (1959), and only a few of the most significant results will be summarised here. The bromination of cholestan-3-one (partial structure (**9.19**)) with molecular bromine in acetic acid containing a little hydrogen bromide gives the 2α-bromo-derivative (**9.20**), whereas that of the isomeric coprostan-3-one (**9.21**) gives 4β-bromocoprostan-3-one (**9.22**). This difference in orientation has been associated with the

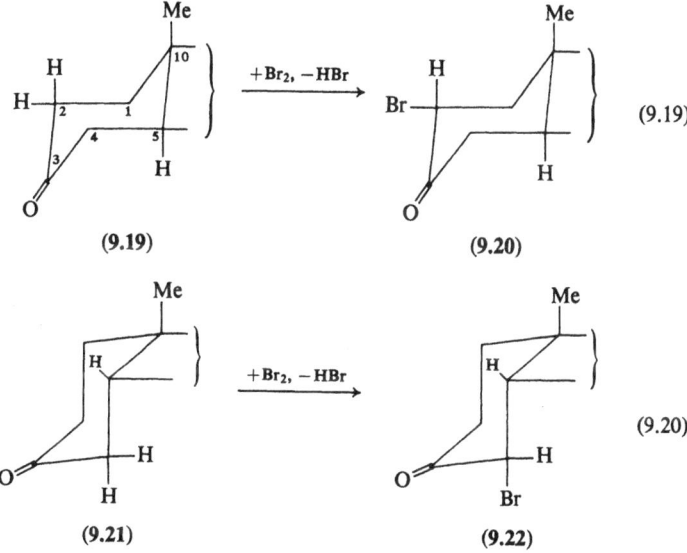

relative ease of enolisation in the two alternative directions; in the cholestane (5α-H) series, the complicated balance of steric interactions favours the formation of the $\Delta^{2,3}$-enol, whereas in the coprostane (5β-H) the $\Delta^{3,4}$-enol is formed more rapidly. In the enol from either series (e.g. 3-hydroxycholest-2-ene; structure (**9.23**) with R = Me) axial attack

(**9.23**)

by halogen on the double bond is expected, and could occur from either face of the molecule. In the absence of other factors, attack on the β-face might have been expected in the cholestane, since the entering bromine could then immediately adopt the axial conformation of the hydrogen atom removed in forming the enol. Steric hindrance to attack on this side of the 2-carbon atom is, however, provided by the substituent at C-10 (R = Me), so instead α-attack predominates as shown in (9.19), giving at first a strained conformation of the bromoketone which then reverts to its most stable form with bromine disposed equatorially.

In the 19-norcholestane series, with a hydrogen instead of a methyl group at C-19 ($\Delta^{2,3}$-enol form; structure (9.23) with R = H) steric hindrance to β-attack is reduced, so that now bromination results in more 2β- than 2α-substitution. Here there is also more enolisation towards the 4-position. The balance between attack on the α- and β-faces can be reversed yet again by the introduction of bulky or dipolar substituents (Me or F) at the 5-position, when again α-substitution predominates. Analogous principles apply to the many other examples of substitutions in steroidal and related cyclic ketones, and much work remains to be done to clarify the extent to which the direction and stereochemistry of halogenation can be controlled.

9.4 Acid-catalysed halogenation of aldehydes

The acid-catalysed halogenations of aldehydes are usually represented as proceeding by mechanisms similar to those operative for ketones, but they deserve separate treatment because of the existence of a number of routes which can compete with reaction paths involving enolisation. When acetaldehyde is treated with chlorine, progressive substitution occurs, to give ultimately chloral ($Cl_3C.CHO$); and similar reactions occur with bromine. Dawson, Burton and Ark (1914) summarised the earlier preparative literature, and showed that for bromination in aqueous solution the acid-catalysed reaction, which was dominant at high acidity, has a rate which is independent of the bromine concentration and is the same as that for iodination. Other aldehydes having an α-hydrogen atom behave similarly, but formaldehyde, with no possibility of enolisation, gave no acid-catalysed reaction. These results have been confirmed by McTigue and Sime (1967), who showed also that the acid-catalysed bromination is subject to the expected reverse solvent deuterium isotope effect ($k_{D_2O}/k_{H_2O} = 2$) except at acidity greater than 6 M H_2SO_4, when bromine began to enter the rate equation and a normal solvent isotope effect would be expected.

It was shown also in Dawson's pioneer investigation that in dilute aqueous solution at acidities below about 0.1 M a second reaction, with kinetic form

$$-d[Br_2]/dt = k[Aldehyde][Br_2]$$

is of dominant importance and that this represents the rate of oxidation of the aldehyde to the corresponding carboxylic acid according to (9.21).

$$CH_3.CHO + Br_2 + H_2O \longrightarrow CH_3.CO_2H + 2HBr \qquad (9.21)$$

Kaplan (1958) established that this reaction is subject to a large primary deuterium isotope effect ($k_{CH_3CHO}/k_{CH_3CDO} = 4$); and Perlmutter-Hayman and Weissmann (1962, 1969) showed that molecular bromine is more effective than hypobromous acid in this oxidation. The oxidations of aldoses to aldonic acids, and of the formate ion to carbon dioxide, (9.22), seem to be related mechanistically.

$$Br_2 + HCO_2^- \longrightarrow H^+ + 2Br^- + CO_2 \qquad (9.22)$$

In most of these investigations, the halogen molecule has been regarded as functioning by electrophilic attack on the aldehydic hydrogen atom, which is removed, (9.23), as hydride ion from the hydrate of the aldehyde **(9.24)**. The resulting hydrogen bromide would then dissociate; and the carbocation **(9.25)**, though heavily stabilised by resonance with the

$$
\begin{array}{cc}
\quad R & \quad R \\
\quad | & \quad | \\
Br\!-\!Br \quad H\!-\!C\!-\!OH \longrightarrow Br^- + Br\!-\!H + {}^+C\!-\!OH & (9.23) \\
\quad | & \quad | \\
\quad OH & \quad OH \\
\quad \textbf{(9.24)} & \quad \textbf{(9.25)}
\end{array}
$$

hydroxyl groups, would lose a proton very rapidly to give the carboxylic acid. This view is consistent with the large primary deuterium isotope effect; with the fact that the corresponding oxidation of ethanol to acetaldehyde is much slower; and with the observation (Cox and Mc-Tigue, 1964) that a pathway catalysed by general bases, regarded as helping to remove the proton from the hydroxyl group, can be established.

This is, however, not the only possible way in which these reactions could be represented. Electrophilic attack by halogen on oxygen is also possible for all these substrates, and Cullis and Swain (1962) have described the similar kinetic term in the slower iodination of acetaldehyde in water in terms of a different type of intermediate, **(9.26)**, involving electrophilic attack not on hydrogen but on oxygen: (9.24). In this addition–elimination sequence, catalysis by general bases, a large

$$\text{Me.CHO} + \text{IOH} \; \rightleftharpoons \; \underset{\underset{(9.26)}{\overset{|}{\text{OH}}}}{\overset{\overset{\text{Me}}{\underset{|}{}}}{\text{H}-\text{C}-\text{O}-\text{I}}} \; \longrightarrow \; \text{H}^+ + \underset{\underset{\text{OH}}{|}}{\overset{\overset{\text{Me}}{|}}{\text{C}}}=\text{O} + \text{I}^- \quad (9.24)$$

primary deuterium isotope effect and a slower rate of reaction for ethanol would also be expected. Whether the relative effectiveness of different reagents (Br_2, BrOH, I_2) is as expected cannot be judged, since it depends on the unknown equilibrium constants for pre-equilibria involving addition to the carbonyl double bond. The writer feels that the exact way in which the halogen functions in these reactions deserves further study.

The oxidation of aromatic aldehydes to acids is kinetically similar (Barker and Dahm, 1965), and is subject to relatively small structural effects represented by a value of $\rho = 0.17$. Replacement of halogen at the aldehydic carbon atom to give the acid chloride instead of the acid can sometimes be carried out, as in the example of (9.25) (Clarke and Taylor, 1944).

$$o\text{-Cl}.C_6H_4.\text{CHO} \; \xrightarrow[\text{at 150 °C}]{+Cl_2, \, -HCl} \; o\text{-Cl}.C_6H_4.\text{COCl} \quad (9.25)$$

The reactions of halogens with acetals are related to those of aldehyde hydrates. Dusseau, Schaafsma and De Boer (1970) have shown that dimethoxymethane requires a trace of hydrogen bromide to allow the formation of the carbocation $CH_2^+(OMe)$ from its reaction with bromine in sulphur dioxide at $-60\,°C$: (9.26) with R = H. For acetone dimethylacetal, (9.26) with R = Me, on the other hand, no catalyst was needed;

$$R_2C(OMe)_2 \; \xrightarrow[-\text{BrOMe}]{+Br_2} \; R_2C^+(OMe) \quad (9.26)$$

in this case the carbocation Me_2C^+ (OMe) is stabilised additionally by the methyl substituents, and can be formed directly. These reactions, and also the oxidation of ethanol, seem best regarded as involving electrophilic attack on saturated oxygen.

9.5   Acid-catalysed halogenation of carboxylic acids and their esters; of acid halides, and of acid anhydrides

The mechanisms available for the acid-catalysed halogenation of ketones appear to be available also for other types of compound having a carbonyl group and an α-hydrogen atom; and the reactions are generally represented in the same way, with enolisation as the rate-controlling step. The reactions of diethyl malonate have been the subject of detailed kinetic investigations, particularly by Bell and his co-workers (Bell, 1973). Acid-catalysed halogenations involve the enol, and the kinetic form, (9.13), in which the halogen enters the rate equation, can be realised. The halogenations of the enols have been estimated to occur at rates considerably less than those of diffusion-controlled encounter between the reagents.

Esters of other dicarboxylic acids and of phenylacetic acid can be brominated by using bromine or chlorinated by using sulphuryl chloride when in either case thionyl chloride is used as the solvent. This first presumably gives the corresponding acid chloride or di-acid chloride. Mechanistic investigation by Kwart and Scalzi (1964) for a series of p-substituted α-methylphenylacetic acids, $Ar.CH(Me).CO_2H$, established that the rate depended partly on the concentration of bromine, and was subject to a rather small primary hydrogen–deuterium isotope effect. From these features and the effect of structure on the rate of reaction, it was concluded that the mechanism of bromination involves the formation of a protonated intermediate, perhaps of structure (**9.27**), which then undergoes internal proton migration when attacked by electrophilic bromine: (9.27). It is possible that mechanisms of this kind will become recognised more generally in the future.

$$\underset{Cl}{\overset{R_2CH}{\diagdown}}CO \underset{-H^+}{\overset{+H^+}{\rightleftharpoons}} \left[\underset{Cl}{\overset{R_2C\cdots H\cdots}{\diagdown}}C\cdots OH \right]^+ \xrightarrow[-2H^+, -Br^-]{+Br_2} \underset{Cl}{\overset{R_2C(Br)}{\diagdown}}C{=}O \qquad (9.27)$$

$$(\textbf{9.27})$$

The well-known Hell–Volhard–Zelinskii method of α-halogenation of carboxylic acids involves treatment of the acid with bromine or chlorine and phosphorus tribromide. The procedure has been reviewed by Sonntag (1953), and is believed to involve the reaction of the enol form of the acyl bromide: sequence (9.28).

$$R_2CH.CO_2H \xrightarrow[-\frac{1}{3}P(OH)_3]{+\frac{1}{3}PBr_3} R_2CH.COBr \longrightarrow$$

$$R_2C{:}C(OH).Br \xrightarrow[-HBr]{+Br_2} R_2C(Br).COBr \qquad (9.28)$$

The bromination of acetic anhydride in acetic acid was investigated mechanistically in outline by Orton and Jones (1912). The rate was reported to be independent of the concentration of bromine and to be catalysed by acids, including Lewis acids. A rate-determining conversion of the anhydride into its enol form followed by bromination was envisaged, and seems to be established by the experiments reported.

Electrophilic oxidations of mandelic acid and its ring-substituted derivatives by bromine water have been studied by Aukett and Barker (1973). Their results indicate the availability of a mechanism in which hypobromous acid is more effective than molecular bromine, unlike reactions involving hydride transfer. The response of rate to change in substituent was small, but in the 'electrophilic' direction ($\rho = -0.43$). The sequence of (9.29) was proposed, in which the decarboxylation is a carbonyl-forming eliminative fragmentation analogous to the carbonyl-forming 1,2-elimination shown in (9.24). These authors discuss a number of related oxidations which may also involve the intermediacy of organic hypohalites.

$$\text{Ph.CH(OH).CO}_2\text{H} \underset{-\text{BrOH, }+\text{H}_2\text{O}}{\overset{+\text{BrOH, }-\text{H}_2\text{O}}{\rightleftharpoons}} \text{Ph.CH-C} \overset{\nearrow \text{O}}{\underset{\searrow \text{O-H}}{}} \overset{-\text{H}^+, -\text{Br}^-}{\underset{-\text{CO}_2}{\longrightarrow}} \text{Ph.CHO}$$

(9.29)

9.6 Acid-catalysed halogenation of aliphatic nitro-compounds

The general nature of the acid-catalysed halogenation of aliphatic nitro-compounds was established by mechanistic studies many years ago, as has been reviewed by Taylor and Baker (1937). The reactions resemble closely those of ketones; under acid catalysis, a proton can shift from an α-carbon atom to a nitro-group to give, (9.30), the neutral tautomer,

$$\text{R.CH}_2\text{.N} \overset{\nearrow \text{O}}{\underset{\searrow \text{O}}{}} \rightleftharpoons \text{R.CH:N} \overset{\nearrow \text{O}}{\underset{\searrow \text{OH}}{}}$$

(9.30)

(9.28)

the nitronic acid **(9.28)**. Acid-catalysed halogenation then proceeds at a rate independent of the concentration and nature of the halogen, and gives the expected product of substitution, R.CH(Br).NO_2, in competition with the side reaction of rearrangement to the hydroxamic acid, R.C(OH):N.OH.

9.7 Base-catalysed halogenations

The mechanisms of base-catalysed halogenations of ketones and related compounds have been reviewed by Bell (1973), and by Ingold (1969). They are closely related to the corresponding acid-catalysed reactions, but there are important differences, since now the electrophile reacts with a carbanion rather than with a neutral enol. The most usual mechanistic sequence is that shown in (9.9), with the formation of the enolate ion

(9.5) as the rate-controlling step, so that the rate of reaction is of the zeroth kinetic order in halogen. The simple enolate ions apparently react with halogen at a rate approximately that of diffusion-controlled encounter. Other mechanistic features which have been thoroughly documented include large conventional primary isotope effects (k_H/k_D often in the region of 7, as expected for a rate-controlling proton transfer); catalysis by external nucleophiles; and intramolecular assistance to enolisation when a nucleophilic group is suitably placed in the reacting ketone (Bell, Earls and Timimi, 1974; Cox and Hutchinson, 1974).

The effects of substituents on the base-catalysed formation of enolate ions are quite different from those for the corresponding acid-catalysed enolisations. Thus halogen substituents strongly increase the rate of base-catalysed proton loss. This has the result that di- and tri-substitution are much more difficult to prevent, and normally occur on the carbon atom involved in the first substitution. So the iodination of acetaldehyde gives a mixture of products, (9.31), complicated also by the decomposition of the resulting di- and tri-iodo-derivatives in the alkaline solution, (9.32) and (9.33).

$$CH_3.CHO \xrightarrow[-HI]{+I_2} I.CH_2.CHO \xrightarrow[-HI]{+I_2} I_2CH.CHO \xrightarrow[-HI]{+I_2} I_3C.CHO \quad (9.31)$$

$$I_2CH.CHO \xrightarrow[-HCO_2^-]{+OH^-} I_2CH_2 \quad (9.32)$$

$$I_3C.CHO \xrightarrow[-HCO_2^-]{+OH^-} I_3CH \quad (9.33)$$

Earlier interpretations of the orientation of substitution in alkyl ketones are, however, probably mistaken. Swain and Dunlap (1972) have shown that the base-catalysed bromination of butan-2-one gives in almost equal proportion the 1- and 3-bromoketones, which then undergo further reactions, the main products of which are indicated in (9.34) and (9.35).

$$CH_3CO.CH_2.CH_3 \quad \begin{array}{c} \xrightarrow[-HBr]{+Br_2} Br.CH_2.CO.CH_2.CH_3 \longrightarrow \\ Br_3CH + HO_2C.CH_2CH_3 \quad (9.34) \\ \\ \xrightarrow[-HBr]{+Br_2} CH_3.CO.CH(Br).CH_3 \longrightarrow \\ Br_3CH + HO_2C.CH(OH).CH_3 \quad (9.35) \end{array}$$

In aqueous solution, in strongly basic conditions, the halogenation of ketones can be shown kinetically to involve attack by the hypohalite ion on the enolate ion (Bartlett, 1934; Knipe and Cox, 1973). The hypobromite ion is more effective than the hypochlorite ion by a large factor, as would be expected for a reaction in which the hypohalite ion acted as an electrophile. A similar rate law and a similar mechanism have been reported by Lii and Miller (1969, 1971) for the base-catalysed halogenations of nitroethane and of phenylacetylene. The participation of a water molecule in the reaction would seem to be needed to allow the effective displacement of O^{2-} from ClO^-.

A number of other reactions of preparative importance involve electrophilic halogenation of carbonyl compounds. Fieser and Fieser (1972) summarise the selective bromination of unsaturated ketones by pyrrolid-2-one hydrotribromide, as illustrated for example in (9.36). The coupling

$$Ph.CH{:}CH.CO.CH_3 \xrightarrow[-C_4H_7NO,\ -2C_4H_7NO.HBr]{+(C_4H_7NO)_3HBr_3} Ph.CH{:}CH.CO.CH_2Br \tag{9.36}$$

reaction between malonic ester or acetoacetic ester and iodine under basic conditions is also well known: (9.37) with $R = Ac$, CO_2Et.

$$R.CH^-.CO_2Et \xrightarrow[-I^-]{+I_2} R.CH(I).CO_2Et \xrightarrow[-I^-]{+R.CH^-.CO_2Et} \begin{array}{l} R.CH.CO_2Et \\ | \\ R.CH.CO_2Et \end{array} \tag{9.37}$$

10 Electrophilic halogenation of some aromatic heterocyclic systems

'See Mystery to Mathematics fly!...' (A. Pope)

10.1 Introduction

Unsaturated heterocyclic systems exist in great profusion; valuable surveys of their reactions have been given by Albert (1968), by Katritzky and Lagowski (1960, 1967), and by others. Lack of space prevents full coverage here of their modes of reaction, even with a restricted set of electrophiles such as the halogens; and particular attention will be given to aromatic heterocycles. The related non-aromatic systems introduce complications which result from the geometry of the ring as well as from the presence of the hetero-atom or atoms, but otherwise can be treated as conventional olefinic compounds behaving as expected from their structure. Additions to D-glucal and to other unsaturated derivatives of carbohydrates fall into this category (See §§8.3, 8.5.3); here the hetero-atom adopts its expected role as an activating and directing substituent (structure (**8.20**)).

In discussing the reactions of aromatic heterocyclic compounds with electrophilic reagents, attention is often focussed from the practical side on the chemical criterion of aromatic character (i.e. the tendency of the system to 'maintain the type' by engaging in a substitution reaction); and from the theoretical side, on the properties of the ground states of the individual molecules, and on what can be deduced from the charge distribution on the various aromatic carbon atoms concerning the expected reactivities at the nuclear positions. Sometimes the localisation energies for substitution at individual carbon atoms are considered instead, a conventional model for the transition state being used. Such considerations have, however, only limited relevance to the reactions which actually happen. There are a number of reasons why this might have been anticipated; particularly for halogenation, in which the electrophilic reagent carries with it a good nucleophile. The hetero-atoms which are normally encountered (oxygen, nitrogen, sulphur) all have lone-pairs

195

of electrons which can co-ordinate with electrophiles, thus at the same time modifying the original heterocyclic system and producing a new potentially electrophilic species. The starting material can itself also potentially help in removal of a proton, either to complete the sought-for substitution or to modify itself by tautomeric change. The opportunities for substitution by mechanisms other than the S_E2 process (e.g. by mechanisms involving addition–elimination, substitution with rearrangement, or substitution by internal delivery of the electrophile) are very great. Some of them have already been exemplified for compounds containing exocyclic hetero-atoms (§§5.5, 5.10, 6.5, 7.3.1, 7.3.5, 7.3.6, 7.4.4, 7.5.3, 7.6, 8.1, 8.4, 8.7); they can become even more significant when the hetero-atoms form part of the aromatic ring.

Some of the factual material in this chapter is derived from an important review by Eisch (1966) who has pointed out many of the mechanistic difficulties and uncertainties which still present a challenge to investigators. His account deserves careful attention (see also §7.6); in areas of controversy, the present discussion indicates the writer's opinions based on what is currently known.

10.2 Activated aromatic heterocyclic compounds

10.2.1 Pyrrole, furan, thiophene and their derivatives. Pyrrole (10.1) achieves considerable aromatic stability through resonance involving the lone-pairs of electrons on the hetero-atom and the aromatic ring, (10.2)–(10.5). The N-hydrogen, therefore, is capable of prototropic

mobility to any of the carbon atoms of the ring; and in the pyrrole molecule, or in any of its neutral N-derivatives, electrons can be made available at any of these positions at the demand of an electrophile.

Pyrrole is very easily polymerised by acid, so electrophilic halogenation must be carried out under very mild, preferably neutral or alkaline, conditions. Chlorination with sulphuryl chloride in ether has been recorded as giving some 2-chloropyrrole (Mazzara and Borgo, 1905), but iodination with iodine in aqueous potassium iodide gives the tetra-iodo-derivative. Mono-iodination has been studied in neutral solution by using some tri-substituted pyrroles (Doak and Corwin, 1949). The kinetic form is as expected (see also chapter 8). From comparison of the rates of iodination of different tri-substituted pyrroles, it was estimated that the 2- is more rapidly attacked than the 3-position by a factor of about 25. This result can be interpreted on the basis that the transition state for 2-substitution resembles the intermediate (resonance structures (**10.6**)–(**10.8**)) in which the carbocationic charge can be distributed by allylic resonance to the 5- as well as to the 3- and 1-positions. For 3-substitution (analogous intermediate, resonance structures (**10.9**), (**10.10**)), only the 2- and 1-positions can readily accommodate the positive charge.

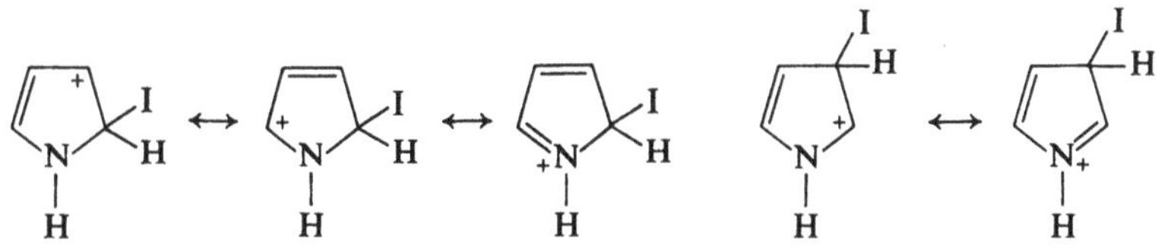

(**10.6**)   (**10.7**)   (**10.8**)   (**10.9**)   (**10.10**)

Methylation of the nitrogen atom increases the rate of iodination by about 10 per cent. This result indicates that halogenation does not require prior ionisation of pyrrole to the conjugate base, or prototropic shift of the hydrogen from nitrogen. Hyperconjugative electron release from the N–H bond is also clearly not important; pyrrole is considered to be planar, or nearly so, and the electrons of the N–H bond are not favourably disposed for overlap with the unsaturation electrons of the ring. The slowness of proton transfer from nitrogen as compared with oxygen may also contribute to the fact that the rate of iodination of pyrrole shows no evidence of the incursion of substitution with rearrangement (the S_E2' mechanism; §7.3), which was evident in aspects of the chemistry of electrophilic attack on phenols.

Furan (**10.11**) behaves similarly, giving first the 2-halogeno- and then the 2,5-dihalogeno-derivatives (e.g. (10.1)).

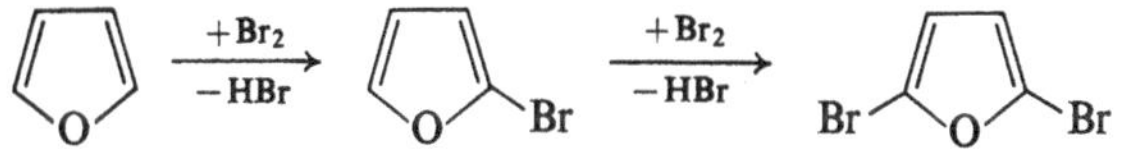

(**10.11**)

The reactions of thiophene (**10.12**) and its derivatives with halogens have been studied kinetically by Marino and his group (Marino, 1965; Linda and Marino, 1967, 1968, 1970; Clementi, Linda and Marino, 1970, 1971) and by Butler and Hendry (1970).

(10.12)    (10.13)

The kinetic forms observed for chlorination and for bromination were similar to those found for other aromatic compounds (chapters 5 and 7). The ratio of 2- to 3-substitution was high (100 or greater), and varied with the reagent over a range of electrophiles in a way which was considered to agree with a linear free-energy treatment, though an anomaly is noted below. Selenophen (10.13) was also investigated; and by including a study of the 2-methoxycarbonyl derivatives of thiophen, furan and pyrrole, and assuming that the effects of substituents are additive in free energies of activation, partial rate factors for 2-bromination were estimated to be for selenophene, 2×10^{11}; for thiophene, 5×10^9; for furan, 6×10^{11}; for pyrrole, 3×10^{18}. Corresponding estimates for benzene activated by lone-pairs borne by exocyclic hetero-atoms are for Ph.OMe, 7×10^9; for Ph.NMe$_2$, 3×10^{19}. The response of rate to change in structure in a series of 5-substituted thiophenes was somewhat lower than for aromatic hydrocarbons and their derivatives.

These reactions are, therefore, thought to be conventional aromatic substitutions: they are not greatly affected by unusual reaction paths or by complexing between the hetero-atom and the electrophilic reagent, though minor changes in the kinetic form were noted by Baciocchi and Mandolini (1968) for the chlorination of benzo[*b*]thiophene. Two discordant observations relating to reaction products deserve further investigation. First, whereas the 2-methoxycarbonyl derivatives of thiophene and of furan give, (10.2), mainly 5-, with very little of the 4-bromo-derivative (see also Chadwick *et al.*, 1973), the corresponding derivative of pyrrole gives, (10.3), 77 per cent of the 4- and 23 per cent of

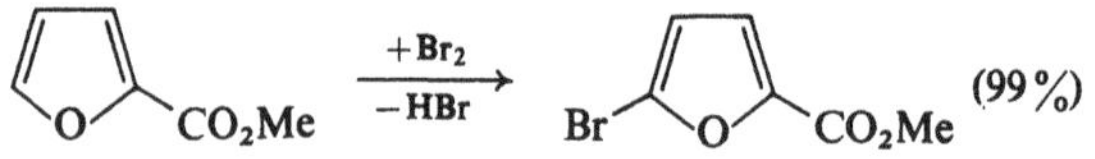

(99%)   (10.2)

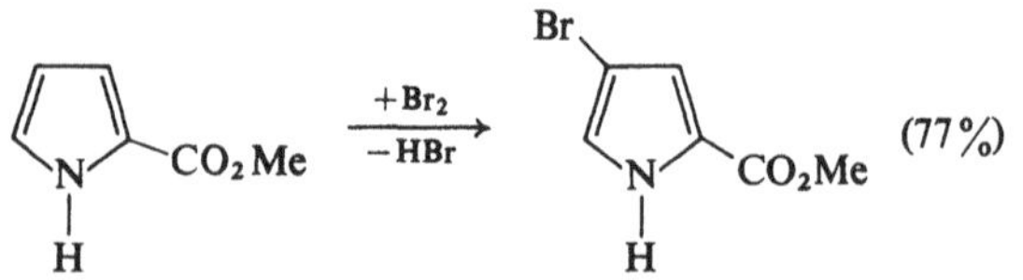

(77%)   (10.3)

the 5-derivative. Secondly, the chlorination of thiophene was found to give more of the 3-derivative with molecular chlorine than with either a source of 'positive bromine' or with molecular bromine. The second of these comparisons is expected from the conventional linear free-energy treatment; the first is not, since molecular chlorine shows a greater response of rate to change of substituent (a higher value of ρ^+) than is found for positive bromination. Whether either or both of these results indicate the incursion of new reaction paths is not known; the incursion of addition–elimination sequences provides possible explanations.

10.2.2 Benzo-derivatives of pyrrole, furan and thiophene. In the benzo-derivatives of these heterocycles, the possibilities for resonance are increased, and electrons can become available on demand at any of the nuclear positions, as is illustrated for benzo[*b*]furan in structures (**10.14**)–(**10.20**). No quantitative estimate of the relative reactivities at the

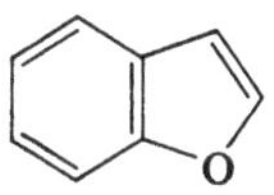

(**10.14**) 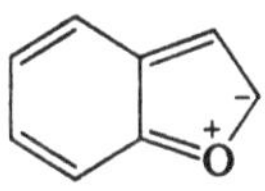   (**10.15**) 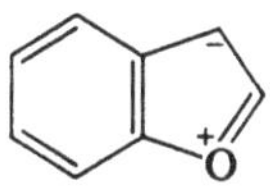   (**10.16**) 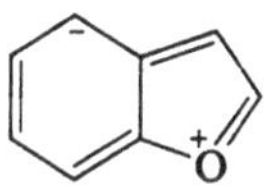   (**10.17**)

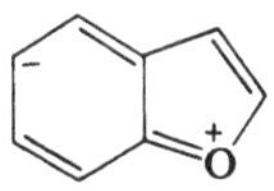

(**10.18**) 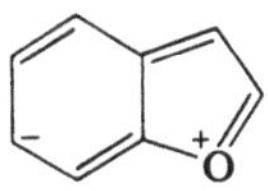   (**10.19**) 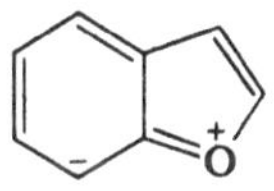   (**10.20**)

various positions can be given; qualitative observations, however, establish that the positions in the heterocyclic ring are considerably more activated than those in the benzenoid ring. Thus indole (benzo[*b*]pyrrole (**10.21**)) gives the 2-chloro-derivative with sulphuryl chloride in ether, but the 3-iodo-derivative with aqueous iodine; benzo[*b*]furan gives with bromine a 2,3-dibromide of unknown stereochemistry; and benzo[*b*]-thiophene (**10.22**) undergoes predominantly 3-bromination and iodination. The fact that 3-substitution is now usually favoured, whereas in the parent heterocycles 2-substitution was preferred, seems reasonable, since activation of the 3-position is possible without interfering with the benzenoid resonance (arrows in structure (**10.22**)). The adjacent benzene

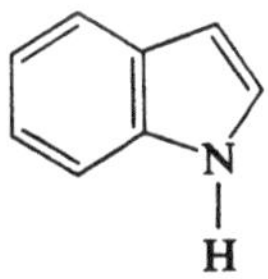

(**10.21**) 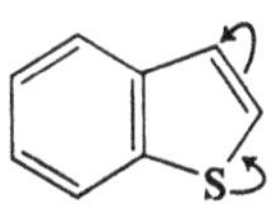   (**10.22**)

ring can be calculated to have increased the reactivity of the 3- and de-creased the reactivity at the 2-position, in each case by a factor which is substantial but small in comparison with the total activation of the system (Clementi, Linda and Marino, 1971).

Activation of the 3-position is sufficiently dominant in indoles that attack at this position may occur even when a blocking 3-alkyl group is present. The resulting carbocation then may take part in a variety of new reactions, the consequences of which have been studied by many groups of workers, and depend for their details on the nature of the electrophile, the conditions of reaction, and the substrate subject to attack. Perhaps the simplest result occurs when the nitrogen atom bears a hydrogen. Then, substitution with double-bond rearrangement (the S_E2' reaction) can give an indolenine, as in (10.4). Gassmann, Campbell and Mehta

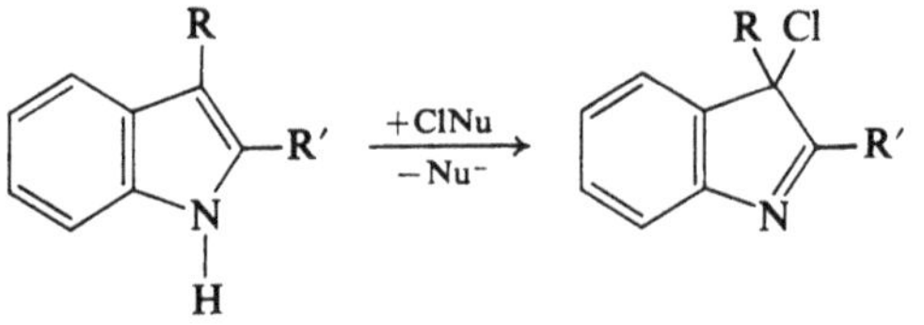

$$\text{(10.4)}$$

(1972) have established that this can happen with R,R′ = Me, Nu = OH, and have discussed the possibility (which they were unable to establish experimentally) that the reaction proceeds through the additional inter-mediacy of the *N*-chloroindole isomeric with the product shown. Other examples clearly analogous are given by Walser, Blount and Fryer (1973) and by Lawson, Patchornik and Witkop (1960).

Another pathway which can follow the initial electrophilic attack leads via addition and oxidation, to an oxindole, as in (10.5) (Hinman and Bauman, 1964). Walser *et al.* (1973) give related examples.

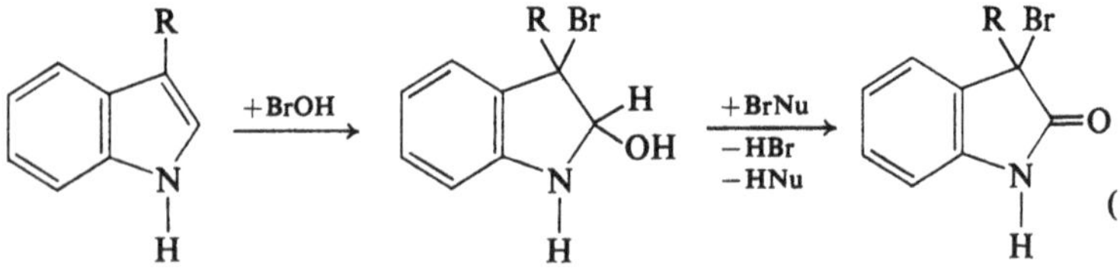

$$\text{(10.5)}$$

In 1-substituted indoles, proton loss from nitrogen is no longer possible; but addition to the 2,3-double bond can still occur. Thus the sequence shown (10.6) has been proposed for the bromination of *N*-acetyl-2,3-dimethylindole, here leading to substitution in the alkyl side chain. In other cases (cf. Da Settimo and Nannipieri, 1970), when the 2-position bears a suitable substituent, an oxindole can be formed.

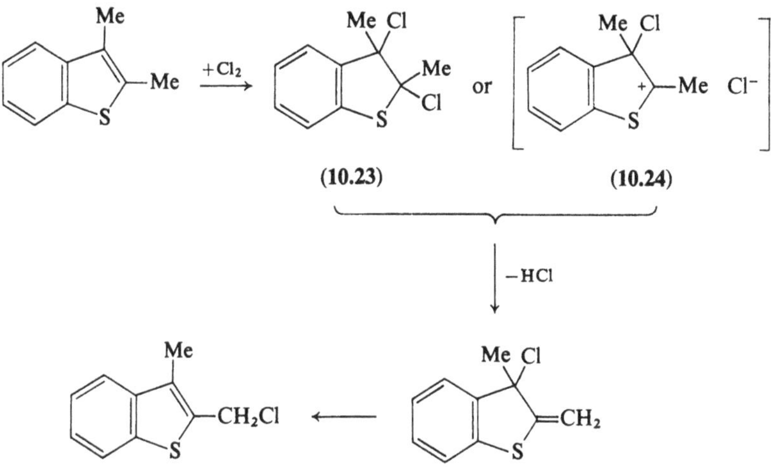

Fig. 10.1. Probable sequence in the chlorination of 2,3-dimethylbenzo[*b*]thiophene.

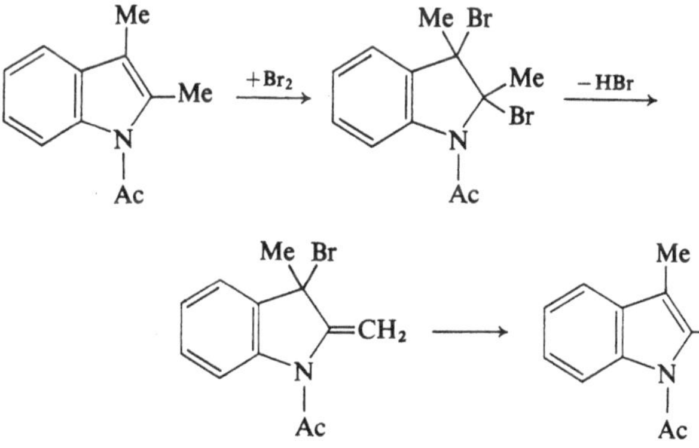

$$(10.6)$$

Substitution in an alkyl side chain has been noted also in 2,3-dimethyl benzo[*b*]thiophene (fig. 10.1; Baciocchi and Mandolini, 1968), where the kinetics and products of chlorination show the quantitative formation of a very stable intermediate, (**10.23**) or (**10.24**), which decomposes quantitatively to give the product of side-chain substitution without the liberation of free chlorine.

10.2.3 Dibenzo-derivatives of pyrrole, furan and thiophene. Substitution in carbazole (**10.25**) and in *N*-acetylcarbazole (**10.26**) are of some interest since partial rate factors for molecular chlorination in acetic acid have been determined and compared with values for non-heterocyclic compounds having related structural features, (**10.27**)–(**10.29**);

(de la Mare, Dusouqui and Johnson, 1966; de la Mare, Wilson and Dusouqui, 1974).

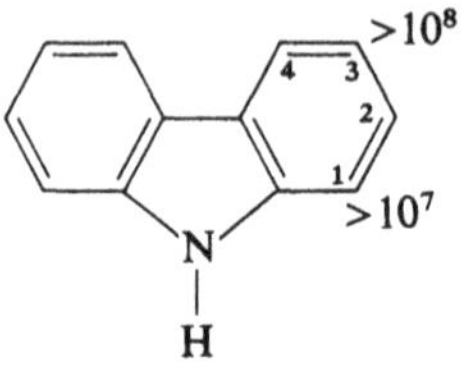

Carbazole itself is powerfully activated for 3- and for 1-substitution, *para* and *ortho* to the nitrogen atom. Acetylation of the nitrogen reduces the reactivity by a large factor, as would be expected. Activation of the 2- and 4-positions, *para* and *ortho* to the second benzene ring (as in fluorene (**10.27**)), now becomes competitive with activation of the 1- and 3-positions, so that all four chloro-*N*-acetylcarbazoles are present in the product of molecular chlorination. When other structural influences are allowed for, only a small influence on the rate can be attributed to the special resonance of the heterocyclic ring; it is a good first approximation to say that each of the positions in the ring has reactivity roughly as expected for a benzene carrying two independently acting substituents.

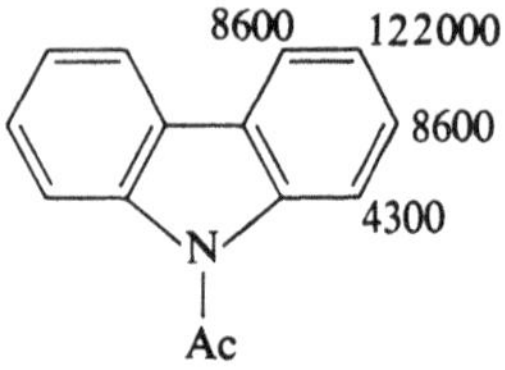

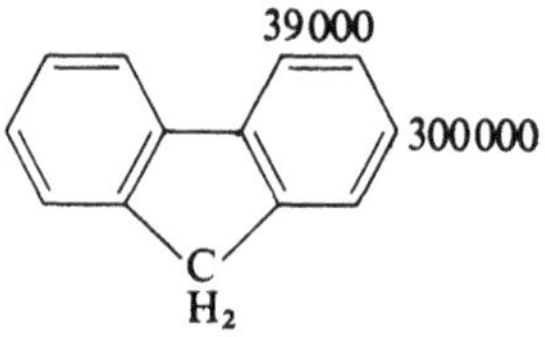

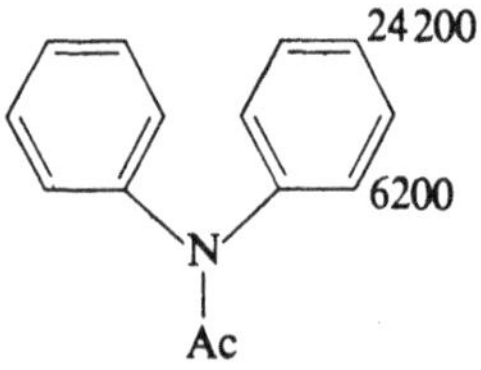

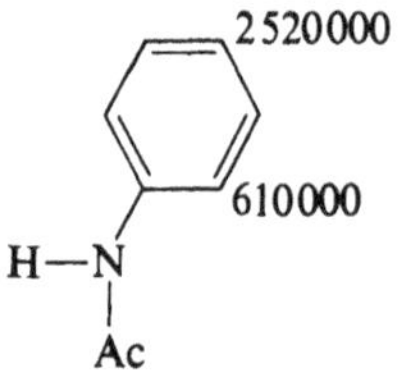

It would be interesting to study dibenzofuran and dibenzothiophene from the same point of view. These compounds have been recorded as undergoing halogenation *para* to the hetero-atom, but further details are not known.

10.2.4 Imidazole, pyrazole, thiazole and related compounds. Imidazole (**10.30**) contains two hetero-atoms; one acts as an activating, and the other as a deactivating substituent for electrophilic substitution. It is convenient to deal with this and related compounds here, because their chemistry on the whole resembles that of other activated heterocycles. Bromination and iodination, (10.7), occur predominantly in the 4-position, which becomes equivalent chemically to the 5-position by

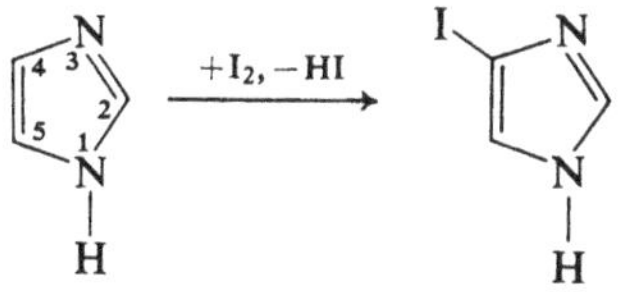

$$\underset{\textbf{(10.30)}}{\text{[imidazole structure]}} \xrightarrow{+\text{I}_2, -\text{HI}} \text{[iodoimidazole structure]} \qquad (10.7)$$

prototropic shift of a hydrogen atom from one nitrogen to the other. Kinetic studies (Grimison and Ridd, 1959) have shown that the reaction involves attack by 'positive iodine' on the imidazole anion, and is subject to a large primary deuterium isotope effect ($k_H/k_D = 4.5$, specific for deuterium in the 4(5)-position). The neutral imidazole molecule is apparently much less reactive (as it should be), since Pauly and Gundermann (1908) showed that N,4-dimethylimidazole does not react with iodine.

Linda (1969) has compared the reactions of imidazole and various of its derivatives with bromine in chloroform. Whereas polysubstitution was observed with imidazole and its N-methyl derivative, benzimidazole underwent N-bromination to form N-bromobenzimidazole hydrobromide (**10.31**), and N-methylbenzimidazole gave only a molecular complex involving the halogen molecule and one or other of the lone-pairs of electrons on nitrogen. It is apparent that replacement of hydrogen on the 2-carbon atom, between the nitrogen atoms, is not favoured in halogenation of imidazole and its derivatives.

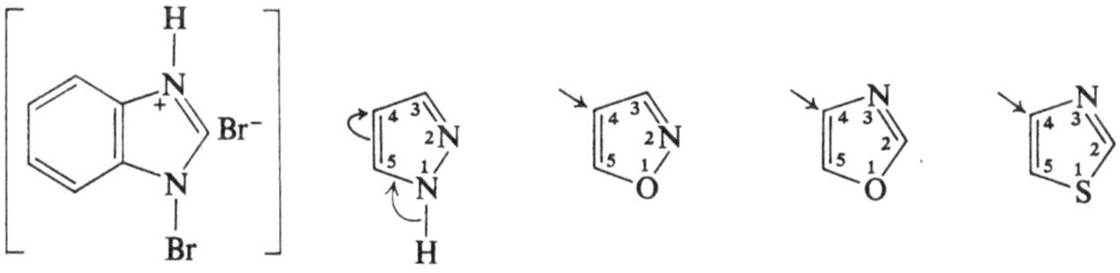

(**10.31**)   (**10.32**)   (**10.33**)   (**10.34**)   (**10.35**)

Pyrazole (**10.32**) is similar to imidazole in many aspects of its behaviour. Both it and its N-methyl-derivative undergo ready 4-halogenation; the arrows in structure (**10.32**) indicate the mode of electron release which favours electrophilic attack on this position. Quantitative studies of relative reactivity are absent for these systems; but general experience indicates that both imidazole and pyrazole are more reactive than benzene, whereas iso-oxazole, oxazole and thiazole (structures (**10.33**), (**10.34**) and (**10.35**) respectively) are less reactive. The expected positions of substitution are as shown by the arrows in the formulae, and the recorded bromination of iso-oxazole accords with this expectation.

10.3   Deactivated aromatic heterocyclic compounds

10.3.1   General and theoretical considerations. The lone-pair of electrons on a pyridine ring has its axis in the plane of the ring; so these electrons are not available for conjugation with the ring, but instead are available for co-ordination with electrophiles. With a strong acid in aqueous solution, an ionised pyridinium salt is formed : (10.8). A more complicated

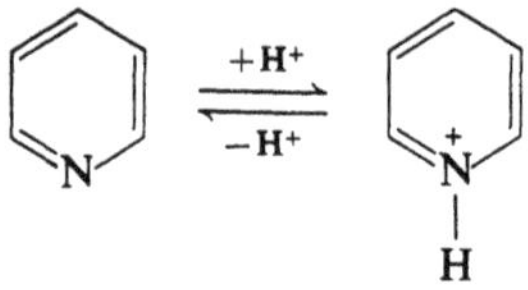

$$\text{(10.8)}$$

sequence, (10.9), might be expected for reaction with a source of electro-philic halogen, particularly in poorly ionising solvents, when complexes and ions, (**10.36**)–(**10.39**), could be produced. Spectroscopic and other evidence for the existence of various complexes of these types has been

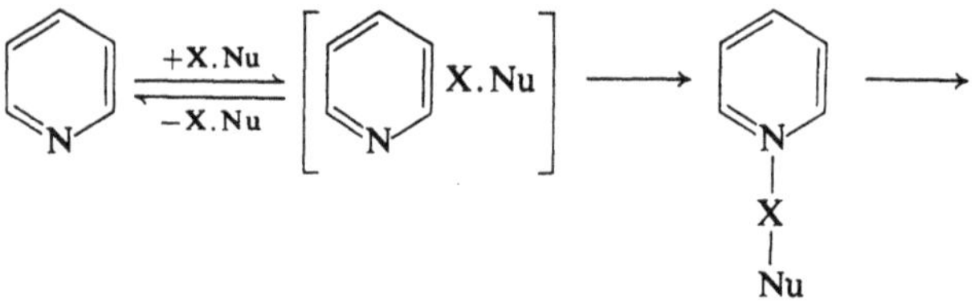

(10.36)
Molecular
complex

(10.37)
σ-Bonded
complex

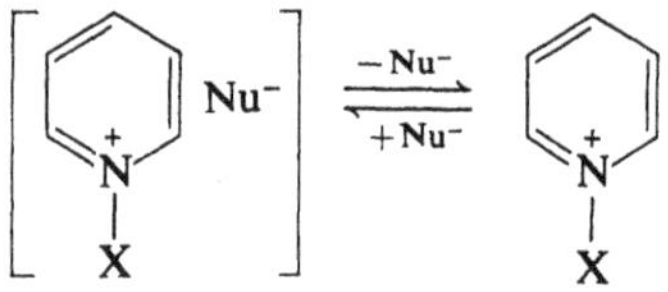

$$\text{(10.9)}$$

(10.38)
Ion-pair

(10.39)
N-Substituted
pyridinium
cation

reviewed by Eisch (1966); the ultimate formation of a *N*-substituted pyridinium ion appears to be paralleled by the formation of adducts from Schiff bases, which contain non-cyclic C=N bonds: (10.10);

$$\text{Ph.CH:NPh} \xrightarrow{\;+\text{Br}_2\;} \text{Ph.}\overset{+}{\text{CH}}.\text{N(Br)Ph} \quad \text{Br}^- \qquad \text{(10.10)}$$

(for a brief review, see de la Mare and Bolton, 1966). It was noted in chapter 2 (§2.7) that cations such as (**10.39**), with X = Br or I, tend to bind a further molecule of halogen, so that salts having formulae [I, 2py]$^+$ClO$_4^-$ and [Br, 2qu]$^+$ClO$_4^-$ (py = pyridine; qu = quinoline) can be characterised.

Prediction of the influence of these reactions on the course of aromatic substitution requires not only knowledge of the equilibrium constants for the formation of the various possible species but also of their intrinsic reactivities with electrophiles and of their rates of formation and de-composition. So far, satisfactory information is almost absent for halo-genation, and theory is only qualitatively helpful. The pyridine nitrogen is a much more strongly basic centre than an aromatic CH group; it is rapidly protonated in aqueous solution. The dissociation constant of its conjugate acid (pK_a = 5.2; cf. dimethylamine, pK_a = 10.8) shows it to be weaker as a base than aliphatic amines by about 5 logarithmic units; this difference reflects the base-weakening influence of the adjacent double bonds, and additionally the greater electronegativity of unsatur-ated nitrogen than of unsaturated carbon. As a result of this, the pyridine molecule would be expected to be deactivated for aromatic substitution; and this deactivation would be expected to be even greater when a positive pole has developed on nitrogen by co-ordination with an electrophile. Even an exocyclic positive pole deactivates an adjacent benzene ring for bromination by 'positive bromine' by a factor of between 10^4 and 10^5 (de la Mare and Hilton, 1962); an even larger inductive influ-ence would be expected when the nitrogen pole is actually within the ring system.

Quantitative extension of these considerations is, however, difficult; and this difficulty is compounded by the fact that molecular orbital calculations on lines which have given sensible results for non-hetero-cyclic systems (chapters 5, 6, 7) have proved to be much less satisfactory for heterocyclic systems. These calculations all predict, whether charge densities or localisation energies are used as a model for the transition state, that the pyridine molecule and the pyridinium cation should be more reactive by many powers of ten than they are found to be (Ridd, 1963). The rate discrepancies are so great that the present writer thinks that it is at present unprofitable to discuss reactivity or orientation of substitution in heterocyclic systems containing unsaturated nitrogen in terms of these theories.

If the reactions considered in (10.9) proceed further to give products of addition to the aromatic ring, an entirely new situation ensues. The products, whether of 1,2- or of 1,4-addition, structures (**10.40**) and

(**10.41**), are cyclic enamines, having nitrogen atoms in which the lone-pairs of electrons are no longer in the plane of the double-bond system,

(**10.40**)   (**10.41**)   (**10.42**)

and so are available for powerful activation for electrophilic attack (see the arrows in structures (**10.40**) and (**10.41**)). Many analogous reactions are known for pyridine and its analogues, and contribute to the tautomeric interconversions so important in the chemistry of these compounds (cf. Albert, 1968; Katritzky and Lagowski, 1960, 1967).

Adducts of this kind are almost certainly important in many halogenations, though at present very little is known to define the details of the sequences in which they become involved. Eisch (1966) has provided an instructive compilation which illustrates how the variety of reaction paths available for quinoline (**10.42**) allows the selection of routes to particular isomeric products. Table 10.1 is derived from his survey, but is expanded to indicate the present writer's views concerning the probable reaction paths involved under the conditions used by different investigators. Further discussion is included in the next section.

10.3.2   Halogenations with sources of 'positive halogen' under strongly acidic conditions. Treatment of quinoline with bromine in sulphuric acid containing silver sulphate at 25 °C was shown by de la Mare, Kiamuddin and Ridd (1960) to result in the formation of 5- and 8-bromoquinoline in approximately equal proportions (table 10.1). Kiamuddin *et al.* (1963, 1964) obtained similar results for the corresponding chlorination and iodination. The orientation resembles that found for nitration by the nitronium ion; it is presumed that the reagent is the halogen cation or a solvated form of this, which attacks the quinolinium cation. The phenanthridinium cation (**10.43**) (Chandler, 1969) has also been brominated by using this method, to give predominantly the 10-bromo-derivative indicated. The pyridinium cation is unreactive; but reaction with pyridine-1-oxide gives in poor yield at high temperature the 2- and 4-bromo-derivatives (Van der Plas *et al.*, 1961), possibly by way of the protonated form (**10.44**) which can be regarded as a *N*-hydroxypyridin-

TABLE 10.1 *Some products of reaction of quinoline (numbering, structure (10.42)) with molecular bromine, and probable mechanisms or reaction paths under the various conditions noted*

Product[a]	Medium	Catalyst	Temp./°C	Yield/per cent	Presumed mechanism or reaction path
1-Dibromide	CCl_4	—	ca 20	100	Co-ordination with N
2-Bromo-	Vapour	—	500	50–60	Free radical (?)
3-Bromo-	CCl_4	Pyridine	77	91	Addition, then dehydrobromination and debromination; or addition, bromination of adduct, then debromination
5-Bromo-	H_2SO_4	Ag_2SO_4	25	28	'Positive bromination' of the quinolinium cation
6-Bromo-	CCl_4	Pyridine	77	ca 2	Addition, bromination of adduct, then bromination
8-Bromo-	H_2SO_4	Ag_2SO_4	25	29	As for 5-bromo-
3,6-Dibromo-	HOAc	—	20	35	Addition, bromination of adduct, then debromination and dehydrobromination
5,8-Dibromo-	H_2SO_4	Ag_2SO_4	25	43	As for 5- and 8-dibromo-
3,6,8-Tribromo-	HOAc	—	20	Not reported	As for 3,6-dibromo-
5,6,8-Tribromo-[b]	H_2SO_4	Ag_2SO_4	25	Good	As for 5- and 8-bromo-

[a] References are as given by Eisch (1966). 4-Bromoquinoline has been reported as a product of heating 3-bromoquinoline hydrobromide at 300 °C; it probably arises through an addition–elimination sequence. Possible conditions conducive to the formation of this and of 7-substituted compounds have been mentioned by de la Mare (1971).

[b] With 3 molecular proportions of bromine and silver sulphate.

(10.43)   (10.44)   (10.45)   (10.46)

ium cation. Isoquinoline when brominated in the presence of excess of aluminium trichloride gives the 5- and 8-bromo-derivatives, probably through the analogous complex (10.46) (Gordon and Pearson, 1964).

These results show qualitatively that the quinolinium cation is de-activated for aromatic bromination by positive bromine, but they do not provide quantitative comparisons with benzene or its derivatives. The only kinetic measurements available (de la Mare, Kiamuddin and Ridd, 1959) are quite incomplete. The crystalline salt having the composition [Br, 2qu]$^+$ClO$_4^-$ was dissolved in water. Its ultraviolet spectrum in solutions of varying acidity were consistent with its dissociation through the equilibria of (10.11) and (10.12). At reasonably high acidity (*ca* 0.5 M)

$$[\text{Br, 2qu}]^+ \ \text{ClO}_4^- \quad \rightleftharpoons \quad [\text{Br, qu}]^+ \ \text{ClO}_4^- + \text{qu} \qquad (10.11)$$

$$[\text{Br, qu}]^+ \ \text{ClO}_4^- + \text{H}_2\text{O} \quad \rightleftharpoons \quad \text{BrOH} + \text{qu} + \text{H}^+ + \text{ClO}_4^- \qquad (10.12)$$

in water at 25 °C, active bromine was found to disappear by an acid-catalysed process having the kinetic form of (10.13), with a rate coefficient based on this equation of 0.02 dm^6 mol^{-2} s^{-1}. About 15 per cent of the

$$-\text{d}[\text{Br}_2]/\text{d}t = k[\text{quH}^+][\text{BrOH}][\text{H}^+] \qquad (10.13)$$

product was a mixture of bromoquinolines, from which the 5-substituted isomer was identified. The nature of the remaining product was not established. At lower acidity, the incursion of another reaction path, anticatalysed by acid, was noted. For benzene under similar conditions, the corresponding rate coefficient is 16 dm^6 mol^{-2} s^{-1} (de la Mare and Hilton, 1962). It follows that the partial rate factor for 5-bromination of the quinolinium cation under these conditions is probably about 10^{-4}, the exact value depending on the assumptions made about that part of the reaction which led to unidentified products. Partial rate factors several powers of ten less than this have been reported by Crout, Penton and Schofield (1971) for the nitration of the quinolinium cation in 80 per cent sulphuric acid. If the reaction inhibited by acid and thus giving a rate minimum at 0.15 M HClO$_4$ is assumed to be associated with bromination of the quinoline molecule, then by using the acidic dissociation constant of the conjugate acid of quinoline it can be estimated that the quinoline is more reactive than the quinolinium cation with hypobromous acid in water by a factor of about 10^4. Since simple products of substitution are minor in amount, these estimates give only upper limits to the rates of the fundamental processes of substitution in the cation and the neutral molecule.

Gilow and Ridd (1974) have examined the rates of bromination of a number of methyl substituted pyridines and 1-methylpyridinium perchlorates with hypobromous acid in aqueous perchloric acid. The kinetic forms indicated that reactions of the pyridines involved electrophilic attack by 'positive bromine' on the pyridinium cations; good yields of the expected products of substitution were obtained, a result which contrasts with that found for quinoline in more weakly acidic solutions. By extrapolating from the effects of substituents on the rate of bromination, the partial rate factor for 3-bromination in the pyridinium cation was estimated to be approximately 6×10^{-13}.

10.3.3 Halogenations involving addition–elimination sequences. Although pyridine and its analogues react only slowly with most electrophilic reagents, a number of their reactions with molecular halogens are relatively rapid. Garcia, Greco and Hunsberger (1960) studied the bromination of pyridine and its derivatives in thionyl chloride at 90 °C. 3,5-Dibromopyridine was formed in good yield; and derivatives gave substitution always β to the ring nitrogen, (10.14), independent of the

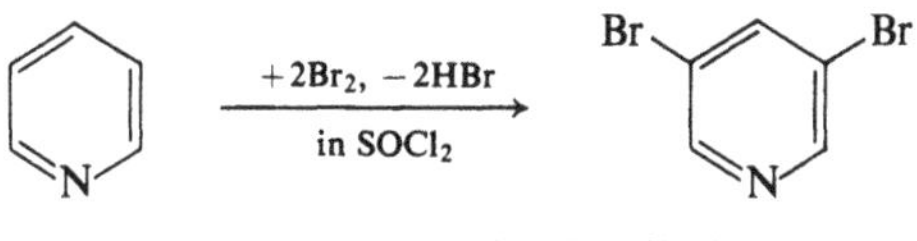

$$\text{(10.14)}$$

polar nature of the substituent. Similarly alkylpyrazines were found to react rapidly, (10.15), with chlorine in carbon tetrachloride to give the 3-substituted product (Hirschberg and Spoerri, 1961).

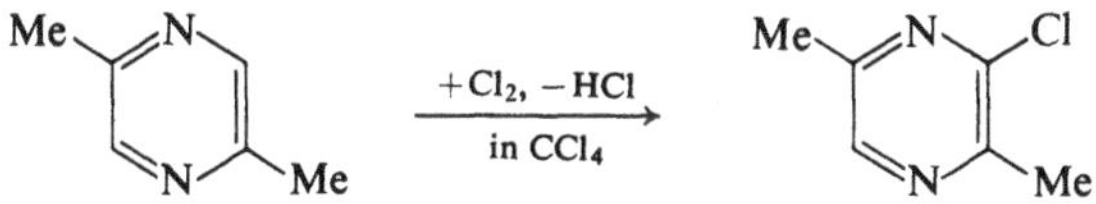

$$\text{(10.15)}$$

Brominations of quinoline (**10.42**) with molecular bromine under various conditions (table 10.1) give complexes involving attachment of electrophilic bromine to nitrogen, which then lose hydrogen bromide to give the 3-bromo-derivative accompanied usually also by di-substituted products of attack on the 6- and 8-positions (Eisch, 1966). The 4-bromination of isoquinoline with molecular bromine (Eisch, 1966) also shows an orientation of substitution different from that which prevails when positively charged reagents attack the isoquinolinium system: (**10.45**) and (**10.46**).

Neither the exact nature of the electrophile nor of the substrate is known for certain with these reactions. The most reasonable hypothesis, however, is that 1,2- or 1,4-adducts are formed; e.g. for pyridine and bromine, structures (**10.40**) or (**10.41**) with X = Nu = Br. These then, even if formed in quite small concentration, would now undergo addition or substitution directed by the lone-pairs of electrons now conjugated with the unsaturated system (see arrows in structures (**10.40**) and (**10.41**)). Dehalogenation and, if necessary, dehydrohalogenation would then give the desired product.

For quinoline, there has been some controversy as to whether 1,2- or 1,4-adducts are involved; and also as to whether an important intermediate on the reaction path is the 1,2,3,4-adduct. de la Mare, Johnson and Ridd (1960; cf. Johnson and Ridd, 1962) have shown that the 1-cyano-quinolinium ion (**10.47**) does not react with bromine in strongly acidic solution, but that reaction is fast in weakly acidic or buffered solution. Under these conditions, a small equilibrium concentration of the 1,2-adduct (the pseudo-base, (**10.48**)) is formed rapidly; this reacts with electrophilic halogenating agents to give adducts (e.g. of structure (**10.49**)), which then can react further with bromine to give products of substitution in the 6- and 8-positions. Decomposition of any of these adducts gives 3-substituted bromoquinolines. Figure 10.2 sets out some of the possibilities. The arrows in structure (**10.49**) show the electronic movements which promote 6- and 8-bromination *para* and *ortho* to the nitrogen atom. Although these results are not directly relevant to the quinoline–bromine reaction, the analogy in pattern of substitution is so close that it seems likely that similar reaction paths are involved in the two processes.

It has been suggested that the bromination of pyridine-*N*-oxide and of quinoline-*N*-oxide with bromine in chloroform containing acetic anhydride and sodium acetate follows the sequence shown in (10.16) (Hamana

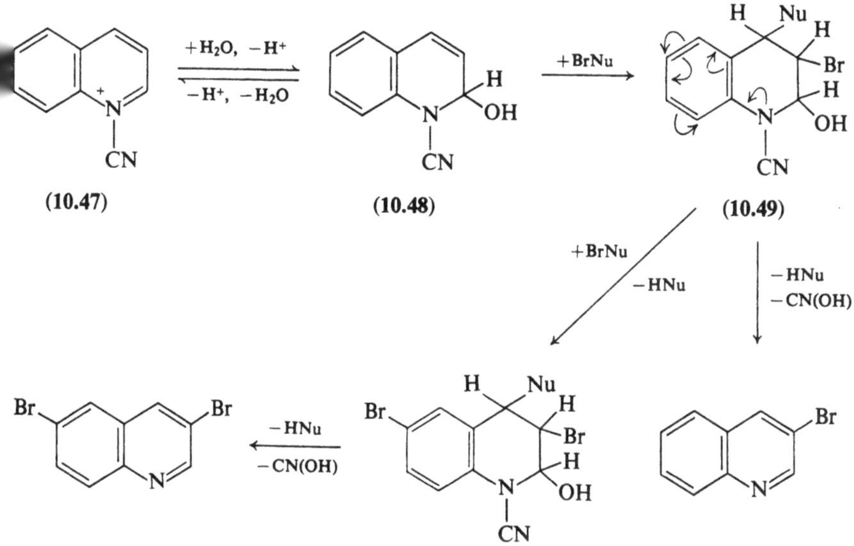

Fig. 10.2. Products of bromination of the *N*-cyanoquinolinium ion in water or in aqueous methanol.

and Yamazaki, 1961). Similarly an adduct can be isolated from the reaction of 2-methyl-5-nitroisocarbostyril with bromine in moist acetic acid, and this decomposes to give 4-bromo-2-methyl-5-nitro-isocarbo-styril: (10.17) (Horning, Lacasse and Muchowski, 1971). Turning to

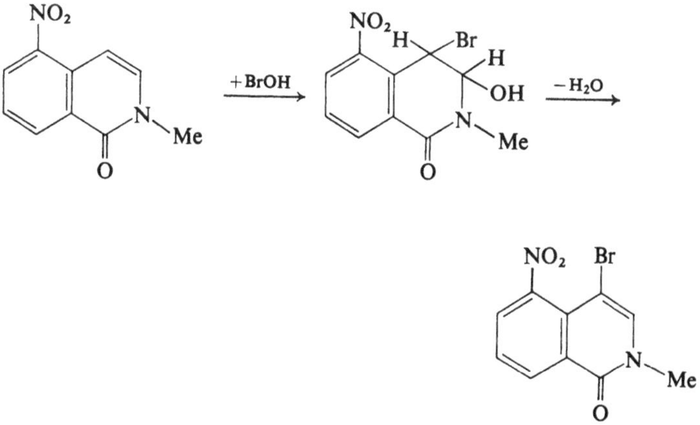

quite a different heterocyclic system, adducts are involved in the 3- and 3,5-dibromination of 2-pyrone: (10.18) (Pirkle and Dines, 1969).

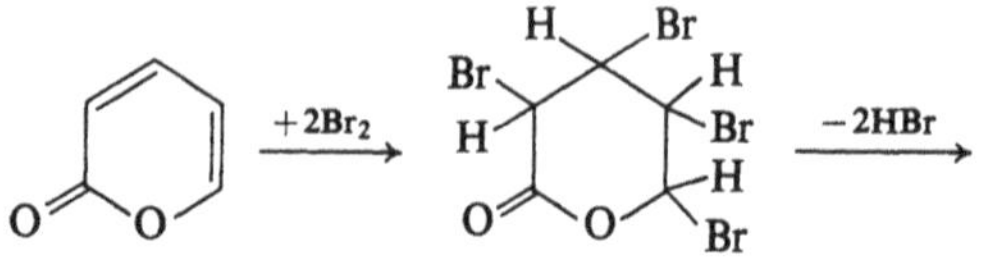

$$(10.18)$$

10.3.4 Halogenations with rearrangement. Many acid-catalysed hydrogen exchanges, and presumably also nitrations, of heterocyclic compounds in strongly acidic solution involve first the attachment of the electrophile at one site and then the removal of a proton from another (Bressel, Katritzky and Lea, 1971). It may be presumed that electrophilic halogenation can proceed in this way also, but no definite cases seem yet to have been established.

Prototropic rearrangement must in some cases precede halogenation of hydroxy- and amino-derivatives of pyridine and its analogues, and of many other heterocyclic systems. It is not, however, essential to the success of these reactions. Brignell, Jones and Katritzky (1970) have investigated kinetically the bromination of several 5-substituted 2-dimethylaminopyridines (structure **(10.50)** with Y = Cl, Br, NO_2) in dilute aqueous solution. The reactions involved the free base, and had the kinetic form usual for these conditions $(-d[Br_2]/dt = k[Br_2][ArH])$. The basicities of the various bases being allowed for, it was shown that the rates were faster than those of the corresponding 5-substituted 2-aminopyridines, and only some three powers of ten slower than those of the corresponding dimethylanilines, **(10.51)**. The writer does not think that this great reactivity, and that of analogous systems, is adequately understood.

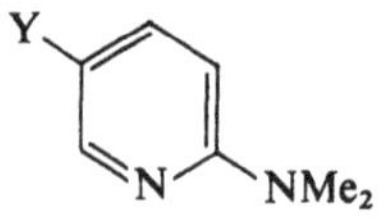

(10.50) 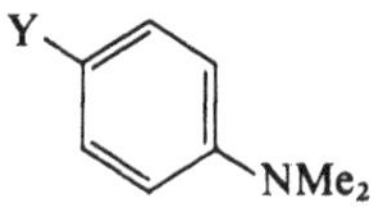   (10.51)

References

Chapter 1

Baker, J. W. (1952). *Hyperconjugation*. Oxford: Oxford University Press.
Bell, R. P. (1973). *The Proton in Chemistry*, second edition. London: Chapman and Hall.
Evans, M. G. and Polanyi, M. (1938). *Trans. Faraday Soc.* **34**, 11.
Fry, A. (1972). *Chem. Soc. Rev.* **1**, 163.
Hammett, L. P. (1938). *Trans. Faraday Soc.* **34**, 156.
Hammett, L. P. (1970). *Physical Organic Chemistry*, second edition. New York: McGraw-Hill.
Hammond, G. S. (1955). *J. Amer. Chem. Soc.* **77**, 334.
Hughes, E. D. and Ingold, C. K. (1935). *J. Chem. Soc.* 244.
Hughes, E. D. and Ingold, C. K. (1952). *Quart. Rev. Chem. Soc.* **6**, 34.
Ingold, C. K. (1969). *Structure and Mechanism in Organic Chemistry*, second edition. London: G. Bell.
Leffler, J. E. (1956). *The Reactive Intermediates of Organic Chemistry*, New York: Interscience.
Leffler, J. E. and Grunwald, E. (1963). *Rates and Equilibria of Organic Reactions*. New York: Wiley.
Melander, L. C. S. (1960). *Isotope Effects on Reaction Rates*. New York: Ronald.
Rochester, C. H. (1970). *Acidity Functions*. London: Academic Press.
Stock, L. M. and Brown, H. C. (1963). *Adv. Phys. Org. Chem.* **1**, 35.
Streitwieser, A., Jr (1962). *Solvolytic Displacement Reactions*. New York: McGraw-Hill.

Chapter 2

Acree, S. F. (1913). *Amer. Chem. J.* **49**, 345.
Arotzky, J. and Symons, M. C. R. (1962). *Quart. Rev. Chem. Soc.* **16**, 282.
Banus, J., Emeléus, H. J. and Haszeldine, R. N. (1951). *J. Chem. Soc.* 60.
Bell, R. P. and Gelles, E. (1951). *J. Chem. Soc.* 2734.
Brønsted, J. N. (1922). *Z. physik. Chem.* **102**, 169.
Brønsted, J. N. (1925). *Z. physik. Chem.* **115**, 337.
Deb, B. M. and Coulson, C. A. (1971). *J. Chem. Soc. (A)*, 958.
Downs, A. J. and Adams, C. J. (1973). *Comprehensive Inorganic Chemistry*, ed. J. C. Bailar, Jr, H. C. Emeléus, R. S. Nyholm and A. F. Trotman-Dickenson, vol. 2, pp. 1107ff. London: Pergamon.
Evans, C. C. and Sugden, S. (1949). *J. Chem. Soc.* 270.
Gillespie, R. J. and Nyholm, R. S. (1958). *Progr. Stereochem.* **2**, 261.

Gillespie, R. J. and Passmore, J. (1972). *Chem. in Britain*, **8**, 475.
Harris, M. M. and Mitchell, R. K. (1960). *J. Chem. Soc.* 1905.
Hassel, O. and Hope, H. (1960). *U.S. Dept. Com. Office Tech. Serv.*, *P.B. Dept. 155139*. (*Chem. Abstr.* **57**, 8020 (1962)).
Hope, H. and Lii, G. H-Y. (1970). *Chem. Comm.* 169.
Lambert, J. B., Johnson, D. H., Keske, R. G. and Mixan, C. E. (1972). *J. Amer. Chem. Soc.* **94**, 8172.
McDaniel, D. H. and Brown, H. C. (1958). *J. Org. Chem.* **23**, 420.
Mauger, R. P. and Soper, F. G. (1946). *J. Chem. Soc.* 71.
Modena, G. and Scorrano, G. (1973). *The Chemistry of the Carbon–Halogen Bond*, ed. S. Patai, part 1, pp. 301–406. London: Wiley.
Orton, K. J. P., Soper, F. G. and Williams, G. (1928). *J. Chem. Soc.* 998.
Parker, A. J. (1962). *Quart. Rev. Chem. Soc.* **16**, 163.
Robertson, H. C., Jr and Acree, S. F. (1915). *J. Amer. Chem. Soc.* **37**, 1902.
Rundle, R. E. (1962). *Records Chem. Progr.* **23**, 195.
Sidgwick, N. V. (1927). *The Electronic Theory of Valency*. Oxford: Oxford University Press.
Sidgwick, N. V. and Powell, H. M. (1940). *Proc. Roy. Soc.* (*A*), **176**, 153.
Stock, L. M. and Brown, H. C. (1963). *Adv. Phys. Org. Chem.* **1**, 35.

Chapter 3

Berliner, E. (1964). *Progr. Phys. Org. Chem.* **2**, 253.
de la Mare, P. B. D. (1971). *Rodd's Chemistry of Carbon Compounds*, second edition, ed. S. Coffey, vol. 3A, pp. 45–88. Amsterdam: Elsevier.
de la Mare, P. B. D. and Bolton, R. (1966). *Electrophilic Additions to Unsaturated Systems*. Amsterdam: Elsevier.
de la Mare, P. B. D. and Swedlund, B. E. (1973). *The Chemistry of the Carbon–Halogen Bond*, ed. S. Patai, part 1, pp. 407–549. London: Wiley.
Eaborn, C. (1960). *Organic Silicon Compounds*. London: Butterworths.
Hartshorn, S. R. (1974). *Chem. Soc. Rev.* **3**, 167.
Huisgen, R., Grashey, R. and Sauer, J. (1964). *The Chemistry of Alkenes*, ed. S. Patai, pp. 806–78. New York: Interscience.
Ingold, C. K. (1969). *Structure and Mechanism in Organic Chemistry*, second edition, pp. 497ff. London: Bell.
Melander, L. (1950). *Arkiv Kemi*, **2**, 211.
Taylor, R. (1972). *Comprehensive Chemical Kinetics*, ed. C. H. Bamford and C. F. H. Tipper, vol. 13, p. 3. Amsterdam: Elsevier.
Williams, D. L. H. (1972). *Comprehensive Chemical Kinetics*, ed. C. H. Bamford and C. F. H. Tipper, vol. 13, pp. 454–61. Amsterdam: Elsevier.
Zollinger, H. (1964). *Adv. Phys. Org. Chem.* **2**, 163.

Chapter 4

Bartlett, N. and Sladky, F. O. (1968). *Chem. Comm.* 1046.
Barton, D. H. R., Ganguly, A. K., Hesse, R. H., Loo, S. N. and Pechet, M. M. (1968). *Chem. Comm.* 806.
Barton, D. H. R., Godinho, L. S., Hesse, R. H. and Pechet, M. M. (1968). *Chem. Comm.* 804.
Barton, D. H. R., Hesse, R. H., Pechet, M. M., Tarzia, G., Toh, H. T. and Westcott, N. D. (1972). *J. Chem. Soc. Chem. Comm.* 122.
Burdon, J., Parsons, I. W. and Tatlow, J. C. (1972). *Tetrahedron*, **28**, 43.
Fahey, R. C. and Schubert, C. (1965). *J. Amer. Chem. Soc.* **87**, 5172.
Grakauskas, V. (1970). *J. Org. Chem.* **35**, 723.

Khutoretskii, V. M., Okhlobystina, L. V. and Fainzil'berg, A. A. (1967). *Russ. Chem. Rev.* **36**, 145.
Merritt, R. F. (1967). *J. Amer. Chem. Soc.* **89**, 609.
Shaw, M. J., Hyman, H. H. and Filler, R. (1969). *J. Amer. Chem. Soc.* **91**, 1563.
Shaw, M. J., Hyman, H. H. and Filler, R. (1970). *J. Amer. Chem. Soc.* **92**, 6498.
Shaw, M. J., Hyman, H. H. and Filler, R. (1971). *J. Org. Chem.* **36**, 2917.
Shaw, M. J., Weil, J. A., Hyman, H. H. and Filler, R. (1970). *J. Amer. Chem. Soc.* **92**, 5096.
Sheppard, W. A. and Sharts, C. M. (1969). *Organic Fluorine Chemistry*, pp. 98ff. New York: Benjamin.
Shieh, T. C., Feit, E. D., Chernick, C. L. and Yang, N. C. (1970). *J. Org. Chem.* **35**, 4020.
Williams, G. H. (1960). *Homolytic Aromatic Substitution*. Oxford: Pergamon.
Zupan, M. and Pollak, A. (1973). *J. Chem. Soc. Chem. Comm.* 845.

Chapter 5

Abrahams, R. J. and Monasterios, J. R. (1973). *J. Chem. Soc. Perkin I*, 1446.
Baciocchi, E., Mandolini, L. and Patara, A. (1975). *Tetrahedron Letters*, 2268.
Bell, F. (1953). *J. Chem. Soc.* 3035.
Bell, F. and Gibson, J. A. (1955). *J. Chem. Soc.* 24.
Bolton, R. (1973). *Comprehensive Chemical Kinetics*, vol. 9, ed. C. H. Bamford and C. F. H. Tipper. Amsterdam: Elsevier.
Burton, G. W., de la Mare, P. B. D. and Wade, M. (1974). *J. Chem. Soc. Perkin II*, 591.
Cabaleiro, M. C. and Johnson, M. D. (1967). *J. Chem. Soc. (B)*, 565.
Cabaleiro, M. C., Johnson, M. D., Swedlund, B. E. and Williams, J. G. (1968). *J. Chem. Soc. (B)*, 1022.
Cristol, S. J., Stermitz, F. R. and Ramey, P. S. (1956). *J. Amer. Chem. Soc.* **78**, 4939.
Datta, R. L. and Bhoumik, J. C. (1921). *J. Amer. Chem. Soc.* **43**, 303.
de la Mare, P. B. D. (1949). *Quart. Rev. Chem. Soc.* **3**, 126.
de la Mare, P. B. D. (1958). *Progr. Stereochem.* **2**, 65.
de la Mare, P. B. D. (1971). *Rodd's Chemistry of Carbon Compounds*, second edition, ed. S. Coffey, pp. 45–88. Amsterdam: Elsevier.
de la Mare, P. B. D. (1974). *Accounts Chem. Research*, **7**, 361.
de la Mare, P. B. D. and Bolton, R. (1966). *Electrophilic Additions to Unsaturated Systems*, pp. 73–108. Amsterdam: Elsevier.
de la Mare, P. B. D. and Ridd, J. H. (1959). *Aromatic Substitution – Nitration and Halogenation*. London: Butterworths.
De Roocker, A. and de Radzitzky, P. (1970). *Bull. soc. chim. Belg.* **79**, 531.
Elion, L. (1923). *Rec. trav. chim.* **42**, 145.
Elion, L. (1925). *Rec. trav. chim.* **44**, 1106, 1108.
Fahey, R. C. (1968). *Topics in Stereochem.* **3**, 237.
Fahey, R. C. and Schubert, C. (1965). *J. Amer. Chem. Soc.* **87**, 5172.
Igarashi, K., Honma, T. and Imagawa, T. (1968). *Tetrahedron Letters*, 755.
Norman, R. O. C. and Taylor, R. (1965). *Electrophilic Substitution in Benzenoid Compounds*, pp. 119–52. Amsterdam: Elsevier.
Poutsma, M. L. (1966). *J. Org. Chem.* **31**, 4167.
Poutsma, M. L. (1968). *J. Org. Chem.* **33**, 4080.
Poutsma, M. L. and Kartch, J. L. (1966). *Tetrahedron*, **22**, 2167.
Taylor, R. (1972). *Comprehensive Chemical Kinetics*, vol. 13, ed. C. H. Bamford and C. F. H. Tipper, pp. 83–135. Amsterdam: Elsevier.
Williams, G. (1941). *Trans. Faraday Soc.* **37**, 749.

Chapter 6

Akiyama, F., Horie, T. and Matsuda, M. (1973). *Bull. soc. chim. Japan*, **46**, 1888.
Andrews, L. J. and Keefer, R. M. (1960). *J. Amer. Chem. Soc.* **82**, 5823.
Baciocchi, E. and Illuminati, G. (1962). *Gazz. Chim. Ital.* **92**, 89.
Boguslavskaya, L. S. (1972). *Russ. Chem. Rev.* **41**, 740.
Bolton, R. (1968). *J. Chem. Soc.* (*B*), 712, 714.
Bolton, R. and de la Mare, P. B. D. (1967). *J. Chem. Soc.* (*B*), 1044.
Burton, G. W., de la Mare, P. B. D., Main, L. and Hannan, B. N. B. (1972). *J. Chem. Soc. Perkin II*, 265.
Caille, S. Y. and Corriu, R. J. P. (1969). *Tetrahedron*, **25**, 2005.
Campaigne, E. and Thompson, W. (1950). *J. Amer. Chem. Soc.* **72**, 629.
Carr, M. D. and England, B. D. (1958). *Proc. Chem. Soc.* 350.
Cotter, J. L., Andrews, L. J. and Keefer, R. M. (1962). *J. Amer. Chem. Soc.* **84**, 793.
de la Mare, P. B. D. and Bolton, R. (1966). *Electrophilic Additions to Unsaturated Systems*. Amsterdam: Elsevier.
de la Mare, P. B. D., Hilton, I. C. and Vernon, C. A. (1960). *J. Chem. Soc.* 4039; de la Mare, P. B. D., Hilton, I. C. and Varma, S. (1960). *J. Chem. Soc.* 4044.
de la Mare, P. B. D. and Main, L. (1971). *J. Chem. Soc.* (*B*), 90.
de la Mare, P. B. D. and Ridd, J. H. (1959). *Aromatic Substitution – Nitration and Halogenation*. London: Butterworths.
de la Mare, P. B. D. and Suzuki, H. (1967). *J. Chem. Soc.* (*C*), 1586.
de la Mare, P. B. D., Wilson, M. A. and Rosser, M. J. (1973). *J. Chem. Soc. Perkin II*, 1480.
Devillier, M. and Bodot, H. (1972). *Bull. soc. chim. France*, 227.
Evans, J. C., Lo, G. Y-S. and Chang, Y. L. (1965). *Spectrochim. Acta*, **21**, 973.
Gassman, P. G. and Campbell, G. A. (1971). *J. Amer. Chem. Soc.* **93**, 2567.
Haberfield, P. and Paul, D. (1965). *J. Amer. Chem. Soc.* **87**, 5502.
Hall, L. D. and Manville, J. F. (1969). *Canad. J. Chem.* **47**, 379.
Hassner, A. and Boerwinkle, F. (1969). *Tetrahedron Letters*, 3309.
Heasley, V. L., Heasley, G. E., Loghry, R. A. and McConnell, M. R. (1972). *J. Org. Chem.* **37**, 2228.
Hüttel, R., Tauchner, P. and Forkl, H. (1972). *Chem. Ber.* **105**, 1.
Kartashov, V. R., Pushkarev, V. P. and Bodrikov, I. V. (1972). *J. Org. Chem. U.S.S.R.* **8**, 23.
Kovacic, P., Lowery, M. K. and Field, K. W. (1970). *Chem. Rev.* **70**, 639.
Kovacic, P. and Sparks, A. K. (1960). *J. Amer. Chem. Soc.* **82**, 5740.
Kovacic, P. and Sparks, A. K. (1961). *J. Org. Chem.* **26**, 1310.
Koyano, T. and Watanabe, O. (1971). *Bull. Chem. Soc. Japan*, **44**, 1378.
Lasne, M.-C., Masson, S. and Thuillier, A. (1972). *Bull. soc. chim. France*, 4592.
Mauger, R. P. and Soper, F. G. (1946). *J. Chem. Soc.* 71.
Neale, R. S., Shepers, R. G. and Walsh, M. R. (1964). *J. Org. Chem.* **29**, 3390.
Streitwieser, A., Jr (1961). *Molecular Orbital Theory for Organic Chemists*. New York: Wiley.
Stroh, R. (1962). *Houben-Weyls Methoden der Organischen Chemie*, ed. E. Muller, vol. 5/3, pp. 760–811. Stuttgart: Georg Thieme.
Traynham, J. G. and Stone, De W. B., Jr (1970). *J. Org. Chem.* **35**, 2025.
Uemura, S., Sasaki, O. and Okano, M. (1971). *Chem. Comm.*, 1064.
Yakubovic, A. Y. and Motsarev, G. V. (1953). *J. Gen. Chem. U.S.S.R.* **23**, 1414.

Chapter 7

Ahmed, S. and Wardell, J. L. (1972). *Tetrahedron Letters*, 2363.

Baciocchi, E. and Illuminati, G. (1967). *J. Amer. Chem. Soc.* **89**, 4017.
Baciocchi, E., Illuminati, G., Sleiter, G. and Stegel, F. (1967). *J. Amer. Chem. Soc.* **89**, 125.
Baliga, B. T. and Bourns, A. N. (1966). *Canad. J. Chem.* **44**, 363, 379.
Barili, P. L., Bellucci, G., Marioni, F., Morelli, I. and Scartoni, V. (1972). *J. Org. Chem.* **37**, 4353.
Beebe, T. R. and Wolfe, J. W. (1970). *J. Org. Chem.* **35**, 2056.
Bellucci, G., Ferretti, M., Ingrosso, G., Marioni, F., Marsili, A. and Morelli, I. (1972). *Tetrahedron Letters*, 3527.
Berliner, E. (1964). *Progr. Phys. Org. Chem.* **2**, 253.
Berliner, E. (1966). *J. Chem. Educ.* **43**, 124.
Berliner, E. and Altschuler, L. (1966). *J. Amer. Chem. Soc.* **88**, 5837.
Berliner, E. and Beckett, M. C. (1957). *J. Amer. Chem. Soc.* **79**, 1425.
Berliner, E., Kim, J. B. and Link, M. (1968). *J. Org. Chem.* **33**, 1160.
Bodrikov, I. V., Bronnikova, N. G. and Okrokova, I. S. (1973). *J. Org. Chem. U.S.S.R.* **9**, 978.
Boguslavskaya, L. S. (1972). *Russ. Chem. Rev.* **41**, 740.
Bradfield, A. E. and Jones, B. (1941). *Trans. Faraday Soc.* **37**, 726.
Buckles, R. E. and Meinhardt, N. A. (1952). *J. Amer. Chem. Soc.* **74**, 1171.
Buckles, R. E. and Womer, W. D. (1958). *J. Amer. Chem. Soc.* **80**, 5058.
Calo, V., Lopez, L., Pesce, G. and Todesco, P. E. (1974). *J. Chem. Soc. Perkin II*, 1192.
Christen, M., Koch, W., Simon, W. and Zollinger, H. (1962). *Helv. chim. acta*, **45**, 2077.
Christen, M. and Zollinger, H. (1962). *Helv. chim. acta*, **45**, 2057, 2066.
Clarke, C. A. and Williams, D. L. H. (1966). *J. Chem. Soc. (B)*, 1126.
Dalton, D. R. and Dutta, V. P. (1971). *J. Chem. Soc. (B)*, 85.
Dalton, D. R., Dutta, V. P. and Jones, D. C. (1968). *J. Amer. Chem. Soc.* **90**, 5498.
Dalton, D. R., Smith, R. C., Jr and Jones, D. G. (1970). *Tetrahedron*, **26**, 575.
Dauphin, G., Kergomard, A. and Scarset, A. (1973). *Bull. soc. chim. France (II)*, 1105.
de la Mare, P. B. D. (1949). *Quart. Rev. Chem. Soc.* **3**, 127.
de la Mare, P. B. D. (1963). *Molecular Rearrangements*, ed. P. DeMayo, pp. 27–110. New York: Interscience.
de la Mare, P. B. D. (1971). *Rodd's Chemistry of Carbon Compounds*, second edition, ed. S. Coffey, pp. 45–88. Amsterdam: Elsevier.
de la Mare, P. B. D. and Bolton, R. (1966). *Electrophilic Additions to Unsaturated Systems*. Amsterdam: Elsevier.
de la Mare, P. B. D. and Dusouqui, O. M. H. El (1967). *J. Chem. Soc. (B)*, 251.
de la Mare, P. B. D. and Hannan, B. N. B. (1973). *J. Chem. Soc. Perkin II*, 1086.
de la Mare, P. B. D. and Lomas, J. S. (1965). *J. Chem. Soc.* 5739.
de la Mare, P. B. D., Scott, R. A. and Robertson, P. W. (1945). *J. Chem. Soc.* 509.
de la Mare, P. B. D. and Singh, A. (1972). *J. Chem. Soc. Perkin II*, 1801.
de la Mare, P. B. D. and Swedlund, B. E. (1973). *The Chemistry of the Carbon–Halogen Bond*, part 1, ed. S. Patai, pp. 407–548. New York: Wiley.
de la Mare, P. B. D. and Wilson, R. D. (1974). *Tetrahedron Letters*, 3777.
Dubois, J. E., Hegarty, A. F. and Bergmann, E. D. (1972). *J. Org. Chem.* **37**, 2218.
Dubois, J. E. and Huynh, X. Q. (1971). *Tetrahedron Letters*, 3369.
Dubois, J. E. and Mouvier, G. (1963). *Tetrahedron Letters*, 1325.
Dubois, J. E. and Mouvier, G. (1964). *Compt. rend. Acad. Sci. Paris*, **259**, 201.
Dubois, J. E. and Mouvier, G. (1965). *Tetrahedron Letters*, 1629.
Dubois, J. E. and Schwarcz, A., (1964). *Tetrahedron Letters*, 2167.
Erickson, K. L. and Barowsky, H. W. (1971). *Chem. Comm.* 1596.
Fahey, R. C. (1968). *Topics Stereochem.* **3**, 237.
Fahey, R. C. and Schneider, H. J. (1968). *J. Amer. Chem. Soc.* **90**, 4429.
Farrell, P. G. and Mason, S. F. (1963). *Nature, Lond.*, **197**, 590.

Fieser, L. F. and Fieser, M. (1967). *Reagents for Organic Synthesis*, vol. 1, pp. 333, 967. New York: Wiley.

Gilow, H. M. and Ridd, J. H. (1973). *J. Chem. Soc. Perkin II*, 1321.

Grob, C. A. and Winstein, S. (1952). *Helv. chim. acta*, **35**, 782.

Hageman, H. J. and Havinga, E. (1966). *Rec. trav. chim.* **85**, 1141.

Hallas, G. and Hepworth, J. D. (1974). *Educ. Chem.* **11**, 25.

Hassner, A. (1971). *Acc. Chem. Res.* **4**, 9.

Hegarty, A. F., Lomas, J. S., Wright, W. V., Bergmann, E. D. and Dubois, J. E. (1972). *J. Org. Chem.* **37**, 2222.

Hughes, E. D. and Ingold, C. K. (1952). *Quart. Rev. Chem. Soc.* **6**, 34.

Huyser, E. S. and Kim, L. (1968). *J. Org. Chem.* **33**, 1243.

Illuminati, G. and Marino, G. (1956). *J. Amer. Chem. Soc.* **78**, 4975.

Israel, G. C., Tuck, A. W. N. and Soper, F. G. (1945). *J. Chem. Soc.* 547.

Kanyaev, N. P. (1961). *Isvest. Vysshikh. Ucbeb Zavendenii, Khim. i. Khim. Tekhnol.* **4**, 225. (*Chem. Abstr.* **55**, 27015).

Kergomard, A. (1961). *Bull. soc. chim. France*, 2360.

Kim, J. B., Chen, C., Krieger, J. K., Judd, K. R., Simpson, C. C. and Berliner, E. (1970). *J. Amer. Chem. Soc.* **92**, 910.

Kirk, D. N. and Hartshorn, M. P. (1968). *Steroid Reaction Mechanisms.* Amsterdam: Elsevier.

Kooyman, E. C. and Louw, R. (1962). *Rec. trav. chim.* **81**, 365.

Lloyd, J. B. F. and Ongley, P. A. (1965). *Tetrahedron*, **21**, 245.

McDonald, I. R. C., Milburn, R. M. and Robertson, P. W. (1950). *J. Chem. Soc.* 2836.

Mach, M. H. and Bunnett, J. F. (1974). *J. Amer. Chem. Soc.* **96**, 936.

Menzel, P. and Effenberger, F. (1972). *Angewandte Chem., Internat. Ed.* **11**, 922.

O'Bara, E. J., Balsley, R. B. and Starer, I. (1970). *J. Org. Chem.* **35**, 16.

Olah, G. A. and Bollinger, J. M. (1968). *J. Amer. Chem. Soc.* **90**, 947.

Parrilli, M., Barone, G., Adinolfi, M. and Mangoni, L. (1973). *Gazz. chim. Ital.* **103**, 1265.

Pearson, D. E., Wysong, R. D. and Breder, C. V. (1967). *J. Org. Chem.* **32**, 2358.

Pfeiffer, P. and Wizinger, R. (1928). *Liebigs Ann.* **461**, 132.

Pincock, J. A. and Yates, K. (1970). *Canad. J. Chem.* **48**, 2944, 3332.

Reich, H. and Lardon, A. (1946). *Helv. chim. acta*, **29**, 671.

Reich, H. J. and Cram, D. J. (1969). *J. Amer. Chem. Soc.* **91**, 3505, 3517, 3527.

Robertson, J. D. and Swedlund, B. E. (1974). *Chem. in New Zealand*, **38**, 129.

Robertson, P. W. (1955). *Sci. Progr.* 418.

Rolston, J. H. and Yates, K. (1969). *J. Amer. Chem. Soc.* **91**, 1469, 1483.

Rothbaum, H. P., Ting, I. and Robertson, P. W. (1948). *J. Chem. Soc.* 980.

Ruasse, M. F. and Dubois, J. E. (1972). *J. Org. Chem.* **37**, 1770.

Scott, J. M. W. and Martin, J. G. (1965). *Canad. J. Chem.* **43**, 732.

Stock, L. M. and Brown, H. C. (1963). *Adv. Phys. Org. Chem.* **1**, 35.

Szmuszkovicz, J. (1963). *Adv. Org. Chem.* **4**, 1.

Taylor, E. C. and McKillop, A. (1970). *Acc. Chem. Res.* **3**, 338.

Taylor, R. (1972). *Comprehensive Chemical Kinetics*, vol. 13, ed. C. H. Bamford and C. F. H. Tipper, pp. 1–388. Amsterdam: Elsevier.

Uemura, S., Sohma, K., Okano, M. and Ichikawa, K. (1971). *Bull. Soc. Chem. Japan*, **44**, 2490.

Williams, D. L. H., Bienvenüe-Goetz, E. and Dubois, J. E. (1969). *J. Chem. Soc. (B)*, 517.

Williams, G. (1941). *Trans. Faraday Soc.* **37**, 749.

Winstein, S. and Shatavsky, M. (1956). *Chem. and Ind.* 56.

Yates, K., McDonald, R. S. and Shapiro, S. A. (1973). *J. Org. Chem.* **38**, 2460.

Ziegler, J. B. and Shabica, A. C. (1952). *J. Amer. Chem. Soc.* **74**, 4891.

Chapter 8

Amaral, L. do and Melo, S. C. (1973). *J. Org. Chem.* **38**, 800.
Andrews, L. J. and Keefer, R. M. (1956). *J. Amer. Chem. Soc.* **78**, 5623.
Arotsky, J., Darby, A. C. and Hamilton, J. B. A. (1968). *J. Chem. Soc. (B)*, 739.
Arotsky, J., Darby, A. C. and Hamilton, J. B. A. (1973). *J. Chem. Soc. Perkin II*, 595.
Barnett, W. E. and Sohn, W. H. (1972). *Tetrahedron Letters*, 1777.
Batts, B. D. and Gold, V. (1964). *J. Chem. Soc.* 5753.
Bell, R. P. and Brown, J. F. (1936). *J. Chem. Soc.* 1520.
Berliner, E. (1960). *J. Amer. Chem. Soc.* **82**, 5435.
Berliner, E. (1964). *Progr. Phys. Org. Chem.* **2**, 253.
Berliner, E. (1966). *J. Chem. Educ.* **43**, 124.
Bowers, A., Denot, E. and Becerra, R. (1960). *J. Amer. Chem. Soc.* **82**, 4007.
Butler, A. R. and Sanderson, A. P. (1974). *J. Chem. Soc. Perkin II*, 1214.
Cambie, R. C., Hayward, R. C., Roberts, J. L. and Rutledge, P. S. (1974). *J. Chem. Soc. Perkin I*, 1120, 1858, 1864.
Chen, E. M., Keefer, R. M. and Andrews, L. J. (1967). *J. Amer. Chem. Soc.* **89**, 428.
Christen, M., Koch, W., Simon, W. and Zollinger, H. (1962). *Helv. chim. acta*, **45**, 2077.
Cunningham, V. L. and Berliner, E. (1974). *J. Org. Chem.* **39**, 3781.
de la Mare, P. B. D. and Bolton, R. (1966). *Electrophilic Additions to Unsaturated Systems*. Amsterdam: Elsevier.
Derbyshire, D. H. and Waters, W. A. (1950). *J. Chem. Soc.* 3694.
Diner, U. E., Worsley, M. and Lown, J. W. (1971). *J. Chem. Soc. (C)*, 3131.
Fahey, R. C. (1968). *Topics Stereochem.* **3**, 237.
Findlay, M. C., Waters, W. L. and Caserio, M. C. (1971). *J. Org. Chem.* **36**, 275.
Grovenstein, E., Jr and Aprahamian, N. S. (1962). *J. Amer. Chem. Soc.* **84**, 212.
Grovenstein, E., Jr, Aprahamian, N. S., Bryan, C. J., Gnanapragasam, N. S., Kilby, D. C., McKelvey, J. M., Jr and Sullivan, R. J. (1973). *J. Amer. Chem. Soc.* **95**, 4261.
Grovenstein, E., Jr and Schmalsteig, F. C. (1967). *J. Amer. Chem. Soc.* **89**, 5084.
Hall, L. D. and Manville, J. F. (1969). *Canad. J. Chem.* **47**, 361.
Hassner, A. (1971). *Acc. Chem. Res.* **4**, 9.
Hassner, A., Hoblitt, R. P., Heathcock, C., Kropp, J. E. and Lorber, M. (1970). *J. Amer. Chem. Soc.* **92**, 1326.
Hayward, R. C. (1973). Ph.D. thesis, University of Auckland, Auckland, New Zealand.
Hunter, W. H. and Budrow, T. T. (1933). *J. Amer. Chem. Soc.* **55**, 2122.
Krespan, C. G. (1962). *J. Org. Chem.* **27**, 1813.
Lown, J. W. and Joshua, A. V. (1973). *J. Chem. Soc. Perkin I*, 2680.
Mauger, E. and Berliner, E. (1972). *J. Amer. Chem. Soc.* **94**, 194.
Miller, L. L., Kujawa, E. P. and Campbell, C. B. (1970). *J. Amer. Chem. Soc.* **92**, 2821.
Miller, S. I. and Noyes, R. M. (1952). *J. Amer. Chem. Soc.* **74**, 3403.
Muizebelt, W. J. (1971). *Tetrahedron Letters*, 4879.
Ogata, Y. and Aoki, K. (1966). *J. Org. Chem.* **31**, 1625.
Ogata, Y. and Nakajima, K. (1964). *Tetrahedron*, **20**, 2751.
Ogata, Y., Urasaki, I. and Ishibashi, T. (1972). *J. Chem. Soc. Perkin I*, 180.
Petrov, A. A. (1940). *J. Gen. Chem. U.S.S.R.* **10**, 819.
Prévost, C. and Lutz, R. (1934). *Compt. rend. acad. sci. Paris*, **198**, 2264.
Schutte, L. and Havinga, E. (1970). *Tetrahedron*, **26**, 2297.
Senkowski, B. Z. and Panson, G. S. (1961). *J. Org. Chem.* **26**, 943.
Staninets, V. I. and Shilov, E. A. (1971). *Russ. Chem. Rev.* **40**, 272.

Suzuki, H. (1971). *Bull. Soc. Chem. Japan*, **44**, 2871.
Tanner, D. D. and Brownlee, B. G. (1966). *J. Amer. Chem. Soc.* **88**, 771.
Taylor, R. (1972). *Comprehensive Chemical Kinetics*, vol. 13, ed. C. E. Bamford and C. F. H. Tipper, pp. 1–406. Amsterdam: Elsevier.
Vainshtein, F. M., Tomilenko, E. I. and Shilov, E. A. (1963). *Kinetica i Kataliz*, **4**, 357.
Williams, D. L. H., Bienvenüe-Goetz, E. and Dubois, J. E. (1969). *J. Chem. Soc. (B)*, 517.
Wilson, C. V., (1957). *Organic Reactions*, **9**, 332.
Wilson, M. H. and Berliner, E. (1971). *J. Amer. Chem. Soc.* **93**, 208.
Woodward, R. B. and Brutcher, F. V., Jr (1958). *J. Amer. Chem. Soc.* **80**, 209.
Zollinger, H. (1964). *Adv. Phys. Org. Chem.* **2**, 163.

Chapter 9

Abraham, M. H. (1973). *Comprehensive Chemical Kinetics*, ed. C. H. Bamford and C. F. H. Tipper, vol. 12, pp. 135–75. Amsterdam: Elsevier.
Aukett, P. and Barker, I. R. L. (1973). *J. Chem. Soc. Perkin II*, 965.
Barker, I. R. L. and Dahm, R. H. (1965). *Chem. Comm.* 194.
Bartlett, P. D. (1934). *J. Amer. Chem. Soc.* **56**, 967.
Bell, R. P. (1973). *The Proton in Chemistry*, pp. 159ff. London: Chapman and Hall.
Bell, R. P. and Cox, B. G. (1971). *J. Chem. Soc. (B)*, 652.
Bell, R. P., Earls, D. W. and Timimi, B. A. (1974). *J. Chem. Soc. Perkin II*, 811.
Bell, R. P. and Page, M. I. (1973). *J. Chem. Soc. Perkin II*, 1681.
Bell, R. P. and Yates, K. (1962). *J. Chem. Soc.* 1927.
Bender, M. L. and Williams, A. (1966). *J. Amer. Chem. Soc.* **88**, 2504.
Cieciuch, R. F. W. and Westheimer, F. H. (1963). *J. Amer. Chem. Soc.* **85**, 2591.
Clarke, H. T. and Taylor, E. R. (1944). *Organic Syntheses*, collective vol. 1, second edition, p. 155. New York: Wiley.
Cox, B. G. and Hutchinson, R. E. J. (1974). *J. Chem. Soc. Perkin II*, 613.
Cox, B. G. and McTigue, P. T. (1964). *J. Chem. Soc.* 3893; *Austral. J. Chem.* **17**, 1210.
Cullis, C. F. and Swain, P. A. (1962). *J. Chem. Soc.* 3348.
Dawson, H. M., Burton, D. and Ark, H. (1914). *J. Chem. Soc.* **105**, 1275.
de la Mare, P. B. D. (1959). *Tetrahedron*, **5**, 107.
de la Mare, P. B. D. and Dusouqui, O. M. H. El (1967). *J. Chem. Soc. (B)*, 251.
de la Mare, P. B. D. and Swedlund, B. E. (1973). *The Chemistry of Carbon–Halogen Bonds*, ed. S. Patai, pp. 526ff. New York: Wiley.
Deno, N. C. and Fishbein, R. (1973). *J. Amer. Chem. Soc.* **95**, 7445.
Dusseau, H. V., Schaafsma, S. E. and De Boer, T. J. (1970). *Rec. trav. chim.* **89**, 535.
Espenson, J. H. and Williams, D. A. (1974). *J. Amer. Chem. Soc.* **96**, 1008.
Fawcett, F. S. (1950). *Chem. Rev.* **47**, 219.
Fieser, L. F. and Fieser, M. (1959). *Steroids*, pp. 280ff. New York: Reinhold.
Fieser, L. F. and Fieser, M. (1972). *Reagents for Organic Synthesis*, vol. 3, p. 240. New York: Wiley.
Fischer, A., Packer, J. and Vaughan, J. (1962). *J. Chem. Soc.* 3318.
Garbisch, E. W., Jr (1965). *J. Org. Chem.* **30**, 2109.
Gaudry, M. and Marquet, A. (1970). *Tetrahedron*, **26**, 5611, 5617.
Gordon, A. J. (1967). *J. Chem. Educ.* **44**, 461.
House, H. O. (1965). *Modern Synthetic Reactions*, pp. 144–64. New York: Benjamin.
Ingold, C. K. (1969). *Structure and Mechanism in Organic Chemistry*, pp. 794ff. London: Bell.
Iriuchijima, S. and Tsuchihashi, G. (1973). *Bull. Chem. Soc. Japan*, **46**, 929.
Jasor, Y., Gaudry, M., Marquet, A. V. and Bettahar, M. (1973). *Bull. soc. chim. France (II)*, 2732.

Kaplan, L. (1958). *J. Amer. Chem. Soc.* **80**, 2639.
Kirk, D. N. and Hartshorn, M. P. (1968). *Steroid Reaction Mechanisms*, pp. 154–68. Amsterdam: Elsevier.
Knipe, A. C. and Cox, B. G. (1973). *J. Chem. Soc. Perkin II*, 1391.
Kosower, E. M., Cole, W. J., Wu, G.-S., Cardy, D. E. and Meisters, G. (1963). *J. Org. Chem.* **28**, 630.
Kwart, H. and Scalzi, F. V. (1964). *J. Amer. Chem. Soc.* **86**, 5496.
Lapworth, A. (1904). *J. Chem. Soc.* **85**, 30.
Lii, R. R. and Miller, S. I. (1969). *J. Amer. Chem. Soc.* **91**, 7524.
Lii, R. R. and Miller, S. I. (1971). *J. Chem. Soc.* (*B*), 2269, 2271.
Mach, M. H. and Bunnett, J. F. (1974). *J. Amer. Chem. Soc.* **96**, 936.
McTigue, P. T. and Sime, J. M. (1967). *Austral. J. Chem.* **20**, 905.
Marshall, D. R. and Roberts, T. R. (1971). *J. Chem. Soc.* (*B*), 797.
Noyes, R. M. and Koros, E. (1971). *Accounts Chem. Res.* **4**, 235.
Orton, K. J. P. and Jones, M. (1912). *J. Chem. Soc.* **101**, 1720.
Perlmutter-Hayman, B. and Weissmann, Y. (1962). *J. Amer. Chem. Soc.* **84**, 2323.
Perlmutter-Hayman, B. and Weissmann, Y. (1969). *J. Amer. Chem. Soc.* **91**, 668.
Rochester, C. H. (1970). *Acidity Functions*, pp. 169–77. London: Academic Press.
Sonntag, N. O. V. (1953). *Chem. Rev.* **52**, 237.
Straus, F., Kollek, L. and Heyn, W. (1930). *Ber. dtsch. chem. Ges.* **63**, 1868, 1884.
Swain, C. G. and Dunlap, R. P. (1972). *J. Amer. Chem. Soc.* **94**, 7204.
Taylor, T. W. J. and Baker, W. (1937). *Sidgwick's Organic Chemistry of Nitrogen*, second edition, pp. 234ff. Oxford: Oxford University Press.
Teo, K. E. and Warnhoff, E. W. (1973). *J. Amer. Chem. Soc.* **95**, 2728.
Toullec, J. and Dubois, J. E. (1973). *Tetrahedron*, **29**, 2859.
Turnbull, J. H. and Wallis, E. S. (1956). *J. Org. Chem.* **21**, 663.
Wyman, D. P. and Kaufman, P. R. (1964). *J. Org. Chem.* **29**, 1956.

Chapter 10

Albert, A. (1968). *Heterocyclic Chemistry*, second edition. London: Athlone.
Baciocchi, E. and Mandolini, L. (1968). *J. Chem. Soc.* (*B*), 397.
Bressel, U., Katritzky, A. R. and Lea, J. R. (1971). *J. Chem. Soc.* (*B*), 4.
Brignell, P. J., Jones, P. E. and Katritzky, A. R. (1970). *J. Chem. Soc.* (*B*), 117.
Butler, A. R. and Hendry, J. B. (1970). *J. Chem. Soc.* (*B*), 848.
Chadwick, D. J., Chambers, J., Meakins, G. D. and Snowdon, R. L. (1973). *J. Chem. Soc. Perkin I*, 1766.
Chandler, G. S. (1969). *Austral. J. Chem.* **22**, 1105.
Clementi, S., Linda, P. and Marino, G. (1970). *J. Chem. Soc.* (*B*), 1153.
Clementi, S., Linda, P. and Marino, G. (1971). *J. Chem. Soc.* (*B*), 79.
Crout, D. H. G., Penton, J. R. and Schofield, K. (1971). *J. Chem. Soc.* (*B*), 1254.
Da Settimo, A. and Nannipieri, E. (1970). *J. Org. Chem.*, **35**, 2546.
de la Mare, P. B. D. (1971). *Rodd's Chemistry of Carbon Compounds*, second edition, ed. S. Coffey, vol. 3A, pp. 84–6. Amsterdam: Elsevier.
de la Mare, P. B. D. and Bolton, R. (1966). *Electrophilic Additions to Unsaturated Systems*, pp. 252ff. Amsterdam: Elsevier.
de la Mare, P. B. D., Dusouqui, O. M. H. el and Johnson, E. A. (1966). *J. Chem. Soc.* (*B*), 521.
de la Mare, P. B. D. and Hilton, I. C. (1962). *J. Chem. Soc.* 997.
de la Mare, P. B. D., Johnson, M. D. and Ridd, J. H. (1960). *Chem. and Ind.* 1505.
de la Mare, P. B. D., Kiamuddin, M. and Ridd, J. H. (1959). *Chem. and Ind.* 727.
de la Mare, P. B. D., Kiamuddin, M. and Ridd, J. H. (1960). *J. Chem. Soc.* 561.
de la Mare, P. B. D., Wilson, M. A. and Dusouqui, O. M. H. El (1974). *J. Chem. Soc. Perkin II*, 634.

Doak, K. W. and Corwin, A. H. (1949). *J. Amer. Chem. Soc.* **71**, 159.

Eisch, J. J. (1966). *Adv. Heterocyclic Chem.* **7**, 1.

Garcia, E. E., Greco, C. V. and Hunsberger, I. M. (1960). *J. Amer. Chem. Soc.* **82**, 4430.

Gassmann, P. G., Campbell, G. A. and Mehta, G. (1972). *Tetrahedron*, **28**, 2749.

Gilow, H. M. and Ridd, J. H. (1974). *J. Org. Chem.* **39**, 3481.

Gordon, M. and Pearson, D. E. (1964). *J. Org. Chem.* **29**, 329.

Grimison, A. and Ridd, J. H. (1959). *J. Chem. Soc.* 3019.

Hamana, M. and Yamazaki, M. (1961). *Chem. Pharm. Bull.* **9**, 414.

Hinman, R. L. and Bauman, C. P. (1964). *J. Org. Chem.* **29**, 1206.

Hirschberg, A. and Spoerri, P. E. (1961). *J. Org. Chem.* **26**, 2356.

Horning, D. E., Lacasse, G. and Muchowski, J. M. (1971). *Canad. J. Chem.* **49**, 2785.

Johnson, M. D. and Ridd, J. H. (1962). *J. Chem. Soc.* 291.

Katritzky, A. R. and Lagowski, J. M. (1960). *Heterocyclic Chemistry.* London: Methuen.

Katritzky, A. R. and Lagowski, J. M. (1967). *The Principles of Heterocyclic Chemistry.* London: Methuen.

Kiamuddin, M. and Choudhury, A. K. (1963). *Chem. and Ind.* 1840.

Kiamuddin, M. and Haque, M. E. (1964). *Chem. and Ind.* 1753.

Lawson, W. B., Patchornik, A. and Witkop, B. (1960). *J. Amer. Chem. Soc.* **82**, 5918.

Linda, P. (1969). *Tetrahedron*, **25**, 3297.

Linda, P. and Marino, G. (1967). *Chem. Comm.* 499.

Linda, P. and Marino, G. (1968). *J. Chem. Soc.* (*B*), 392.

Linda, P. and Marino, G. (1970). *J. Chem. Soc.* (*B*), 43.

Marino, G. (1965). *Tetrahedron*, **21**, 843.

Mazzara, G. and Borgo, A. (1905). *Gazz. chim. Ital.* **35**, 477.

Pauly, H. and Gundermann, K. (1908). *Ber. dtsch. chem. Ges.* **41**, 3999.

Pirkle, W. H. and Dines, M. (1969). *J. Org. Chem.* **34**, 2239; *J. Heterocyclic Chem.* **6**, 313.

Plant, S. G. P. and Tomlinson, M. L. (1933). *J. Chem. Soc.* 955.

Ridd, J. H. (1963). *Physical Methods in Heterocyclic Chemistry*, **1**, 109.

Van der Plas, H. C., den Hartog, H. J., Van Ammers, M. and Haase, B. (1961). *Tetrahedron Letters*, 32.

Walser, A., Blount, J. F. and Fryer, R. I. (1973). *J. Org. Chem.* **38**, 3077.

Index

acenaphthylene, chlorination of, 76, 80, 102–4, 105
acetaldehyde, 188–90, 193
acetals, 190
acetamidonaphthalenes, fluorination of, 59–60
acetamido substituent, 67–8, 71
acetanilide, and derivatives, 37, 67–8, 118
acetic acid, as solvent, 30, 35, 37, 56, 65ff, 88–90, 91, 102, 103, 105, 119, 127–8, 131ff, 142, 148, 150, 155, 161, 162, 167, 175, 178, 187, 191, 201–2, 207
acetic anhydride, 191–2
    as solvent, 51
acetone, 179–80, 183–4
acetonitrile, as solvent, 169
acetophenones, substituted, 185
3-acetoxycholesta-3,5-diene, 152
acetoxyl, as substituent, 40
acetoxylation, by addition–elimination, 51
N-acetyl carbazole, 201–2
N-acetyl-2,3-dimethylindole, 200
acetylenic compounds, 92–3, 114–15, 116, 133, 169–71
acetyl hypochlorite, see chlorine acetate
acidity functions, 10, 107, 125, 181
acids
    as catalysts, 9, 31, 66, 107–10, 134, 136–7, 179ff, 206ff
    general, see general acids
    Lewis, see Lewis acids
acrylic acid, and derivatives, 67, 172
acrylonitrile, 120

activated processes, 3
activation, free-energy of, 14–15
    see also energy of activation
active hydrogen, 177ff
addition, definition of, 17
addition–elimination pathways, 51–2, 54, 91–2, 155–8, 209–12
aldehydes, halogenation of, 188–90
aldoses, oxidation of, 189
alkyl and aryl migrations, 50–1, 86–7, 122, 148–9, 156–8, 163–4, 174
alkyl aryl sulphenate esters, 178
alkyl aryl sulphides, 157, 178
allene and methyl-substituted derivatives, 43, 93–4, 114–15
allenic cations, 149, 162
allyl bromide, and derivatives, 51, 144–5, 147, 163
allyl chloride, and derivatives, 81, 111–13, 144–5, 147, 163
allyl derivatives, halogenation of, 125, 144–5, 147, 163
allylic rearrangement, 73, 156–7
allyl trimethylammonium perchlorate, 125
aluminium chloride, 208
    as catalyst for additions, 114–15
    see also Lewis acids
amines, aromatic, see aniline
aminopyridines, 212
anchimeric effect, 40
    see also neighbouring-group interaction
aniline, and derivatives, 5ff, 118–19, 153, 156–8, 164–5, 169
    see also acetanilide; dimethylaniline

223

01 14

For EU product safety concerns, contact us at Calle de José Abascal, 56–1°, 28003 Madrid, Spain or eugpsr@cambridge.org.

www.ingramcontent.com/pod-product-compliance
Ingram Content Group UK Ltd.
Pitfield, Milton Keynes, MK11 3LW, UK
UKHW042213151225
466003UK00003B/55